AF789098

GEOLOGY

A PRACTICAL INTRODUCTION FOR SURVEYORS

GARETH W EVANS

2003

EG BOOKS

A division of Reed Business Information

Estates Gazette
151 Wardour Street, London W1F 8BN

© Gareth Evans, 2004

ISBN 0 7282 0423 1

Apart from any fair dealing for the purposes of research or private study, or criticism or review, as permitted under the UK Copyright Designs and Patents Act 1988, this publication may not be reproduced, stored, or transmitted, in any form or by any means, without the prior written permission of the Publishers, or in the case of reprographic reproduction, only in accordance with the terms of the licences issued by the Copyright Licensing Agency in the UK, or in accordance with the terms of licences issued by the appropriate Reproduction Rights Organisation outside the UK. Enquiries concerning reproduction outside the terms stated here should be sent to the Publishers.

The Publishers make no representation, express or implied, with regard to the accuracy of the information contained in this publication and cannot accept any legal responsibility or liability for any errors or omissions.

The material contained in this publication constitutes general guidelines only and does not represent to be advice on any particular matter. No reader or purchaser should act on the basis of material contained in this publication without first taking professional advice appropriate to their particular circumstances. The Publishers expressly disclaim any liability to any person who acts in reliance on the contents of this publication. Readers of this publication should be aware that only Acts of Parliament and Statutory Instruments have the force of law and that only courts can authoritatively interpret the law.

Copying in any form (including all hard copy and electronic formats) is strictly forbidden without written permission from the Publishers, Estates Gazette, a division of Reed Business Information.

Design and typesetting by Ted Masters

Printed in Malta by Interprint Limited

Contents

Tables and Figures

Index of tables

Index of figures

Note and acknowledgements

In this small publication, which leans largely on the backs of professional geologists and engineers, massive simplification is necessary but some useful references are listed at the back for those who wish, or need, to enquire further before seeking specialist help.

I have assumed geological conditions found typically on land in the UK. Occasionally, there is no British example to illustrate a point and I have had to refer to overseas locations. Similarly, a few photographs are of non-British locations. I apologise to all professional geologists, who will find this book excessively condensed; and to those readers who feel that their part of Britain has been left out. Any inconsistencies, errors or omissions are my own. Finally, I have broadened the scope of the book to include the geohazards of flooding, mine subsidence and the like that are of direct relevance.

The glossary at the end of the book, while not encyclopaedic, contains a number of additional terms that are not within the text.

I have received specific help from several individuals and organisations including:

Linda Danischewsky of Salt Union
Dr Clive Edmonds of Peter Brett Associates
Colin McCreath of Arup Scotland, and colleagues
Sean Cripps of Warings

Nearly all the photographs are my own. The others are allowed to be reproduced unattributably. The diagrams have either been re-drawn by compilation from various sources or are my own.

Any mistakes are also my own.

Gareth W Evans, November 2003

Chapter 1 Introduction

What is geology?

Geology, or rather its subject-matter, underlies all of man's activities (Fig.1). Geology is the science of the earth, including the composition, structure and origin of its rocks. Like surveying it draws upon many disciplines, ranging in this case from physics to biology.

The Law

Fig.1 Foundations are not usually exposed, but this cottage is unmistakably built off solid rock.

Law reports underline the significance of geology.

In the case of *Cormack* v *Washbourne* [1996] EGCS 196 (a case that subsequently became more significant in terms of insurance law), a surveyor was criticised for not taking account of a geological map. It was held that he should have recognised that the 'mottled clay' shown on the Geological survey map indicated that there was a risk to the subject property from nearby trees.

However, the situation is not simple as the court supposed, as this handbook will demonstrate.

In *Holbeck Hall Hotel Ltd* v *Scarborough Borough Council* [2000] EGCS 29, the Borough Council escaped blame for a massive (and inevitable) landslip.

In the case *Eley* v *King & Chasemore* [1989] 1 EGLR 181, the surveyor was shown to have given wise advice about shrinkable clay under a property in Burgess Hill, which is a known shrinkable clay area.

Marder v *Sautelle & Hicks* [1988] 2 EGLR 187 relates to the survey of a Cornish bungalow that contained mundic blocks (made from mine waste – see below and glossary).

The surveyor will, in general, be concerned with those aspects that affect the stability, serviceability and security of a building or facility. Subsidence, settlement, landslip, and flood risk may come to mind, as may radon gas.

> **In September, 2000**
> **The Observer quoted the**
> **Subsidence Claims Advisory**
> **Bureau as saying:**
> *'In Britain a lot of surveyors don't really know what subsidence is'*

Existing guidance for surveyors

The RICS 'Red Book' refers to site stability and has a quite a long section on Old Mine Entries. The Guidance Notes for surveys refer only briefly to environmental factors that may be discovered by post-inspection enquiry, and the HSV Guidance Notes contain nothing specific. Site investigation for new construction is normally carried out by engineers. PPG 14 contains a useful digest of subsidence hazards, but will probably not be read by many surveyors. The surveyor can be forgiven, therefore, for supposing that geology is of little relevance.

A new approach

In order to appreciate these risks fully, a brief grounding in geology is necessary. This includes the gross physical characteristics of common rock types; how the structure (attitude) of the rocks affects their performance; and how to derive this information from geological maps. Knowledge of the earth's history and of mining and quarrying is also useful.

Many modern geological techniques (particularly dating) rely upon expensive and complex laboratory techniques which will not generally be relevant to the surveyor. However, basic geological knowledge, the ability to read the geological map, and the naked eye in the field, can tell the surveyor much and permit him/her to interpret environmental search results, etc, that may be unnecessarily negative.

This book serves as a guide or set of signposts to the surveyor of buildings and land, so that he/she may know how and where to seek further guidance. It cannot be relied upon in any given case.

What lies under a building?

Soil

The surface of the ground is usually formed by soil: a topsoil with some organic content and an underlying subsoil of more or less weathered or altered rock. Buildings should not be constructed (although some have been) off the topsoil. Topsoil is not generally the subject of geology. The underlying rock is.

Rock

What is a rock?

Rock is a mixture of minerals. Minerals are difficult to define, but are generally taken to be substances forming the earth that have definite chemical compositions and atomic structure and are formed by inorganic natural processes. Materials such as coal – which is organic – are more difficult to categorise, but present no real problems in practice.

A rock can be:

very soft – unconsolidated silt or clay.
very hard – granite.

Rocks are conventionally divided into:

- **Igneous rocks:** crystallised from molten material underground or as lava flows. More well-known types include granite, dolerite and the popular 'Norway Porphyry' (sold under a variety of trade names such as 'Blue Pearl') found in shop fronts and claddings.

- **Sedimentary rocks:** derived from other rocks or biological activity and deposited usually by wind or water action. Includes clay (such as London Clay), sandstone (such as so-called 'York Stone' beloved of estate agents), shale, and limestones (such as Bath Stone), together with some less frequently encountered rock types.

- **Metamorphic rocks:** crystallised by heat or pressure and changed from their original sedimentary or igneous characters. Slate and marble are the types with household names.

Wide variations occur within each broad grouping. In reality, there is virtually a continuous gradation from fresh sediment to lava.

Detailed rock classification by petrologists is complex and precise identification may depend on laboratory techniques. The surveyor will need to recognise only the more common types for most purposes. These are discussed later at more length. First, however, some background is needed.

Chapter 2 The earth and its history

Introduction

The earth is very active. Hills are impermanent, continents change shape. The crust is a mere skin about 5-60 km thick representing about 1% of the earth's volume. It is broken into sections called plates as shown in Fig.2. The upper view shows the present-day distribution of plate boundaries.

As the crust moves around, like croutons densely covering a bowl of soup, the various plates are in continuous interaction as they collide, separate or slide past each other. Parts of the crust, composed of lighter rocks, form the continents. These do not usually sink and when they collide they raise mountains. The heavier and thinner parts of the crust, which have only a thin layer of sediment on top, underlie the oceans and can sink under the edges of other plates (subduction). This mechanism, termed plate tectonics, became standard theory at the beginning of the 1970s and seems satisfactorily to explain most observed features. The driving force, however, still remains a puzzle. Fig.2 (lower view) shows a 'snapshot' of the earth's past geography based on recent thinking.

As the plates are heaved up at collision areas they are intruded, or pushed into, by molten rock and pierced by volcanoes. The uplifted land is subjected to erosion, which creates sediment. The sediment from the mountains is washed by rivers into lakes and seas, to become compacted into new rock, which is later uplifted as new mountains, and so on.

For many purposes the surveyor will not need to consider the age of the rocks that underlie his/her subject building. However, some understanding is necessary in order to interpret the geological map.

Authorities generally agree that the earth was formed about 4.5 thousand million (4.5 billion) years ago or a little earlier. The oldest dated earth materials are crystal grains (Zircons) about 4.1 billion years old. Life forms developed perhaps about 3.8 billion years ago according to some scientists, but did not appear to form complex or

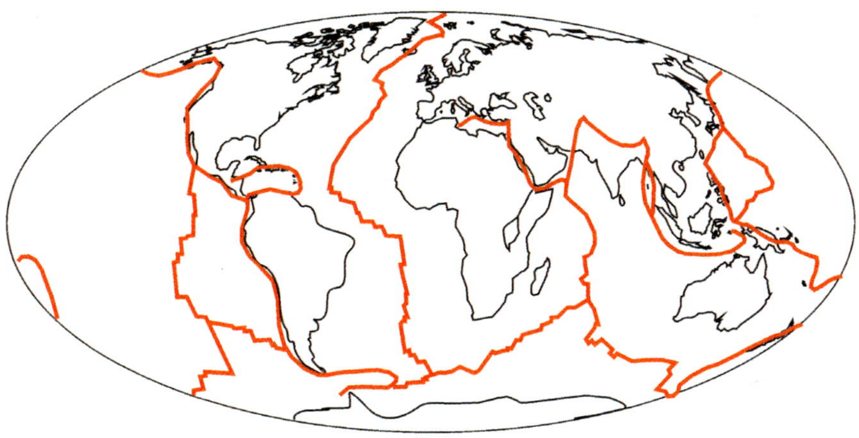

The world showing major plate boundaries (red lines)

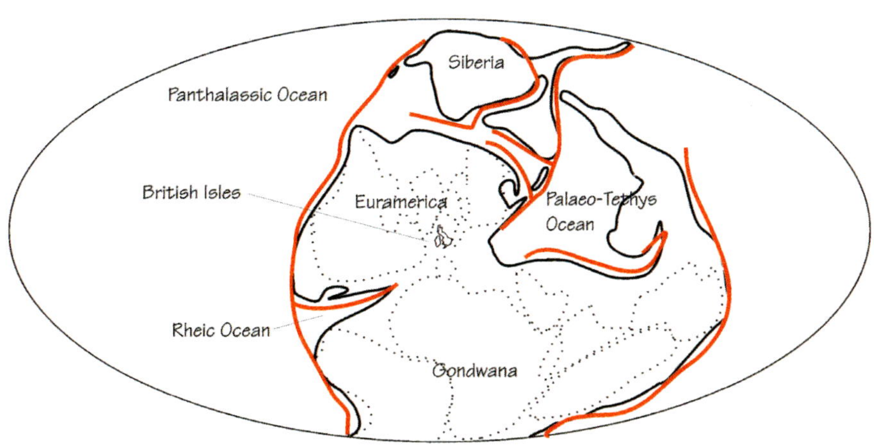

The world in early Carboniferous times (about 360 million years ago) showing major plate boundaries (red lines) and the approximate shapes of modern continents (broken lines). The solid black lines are boundaries of continental crust, not of land masses

Fig.2 Modern (above) and Carboniferous (below) arrangements of crustal plates.

larger organisms until about 580 million years ago (all dates are being constantly revised!).

The earth's interior

It is a popular misconception in some quarters that below the crust, the earth is completely molten internally. In fact, the current thinking is that pockets of molten rock (see discussion below) are found only in limited areas.

As the interior of the earth is naturally inaccessible to us, its structure has been deduced over a long period from the study of seismograms. These are the traces recorded by seismographs, the instruments that detect earthquakes. The different parts of the earth have different densities, which transmit seismic shock waves at different speeds and refract (or bend) the waves in different rays, as a lens refracts light.

At present, geophysicists estimate that the crust has an average thickness of about 38 km. It is, however, much thicker under great mountain ranges, much thinner under rift valleys, and thinner still under the ocean (Fig.3). Below this, there is the mantle, which is thought to consist of magnesium and iron silicates. This extends to a depth of some 2,900 km. Below this, there is the core. The outer core is thought to consist of liquid iron with nickel, oxygen and sulphur impurities. The innermost core is thought to be of solid iron.

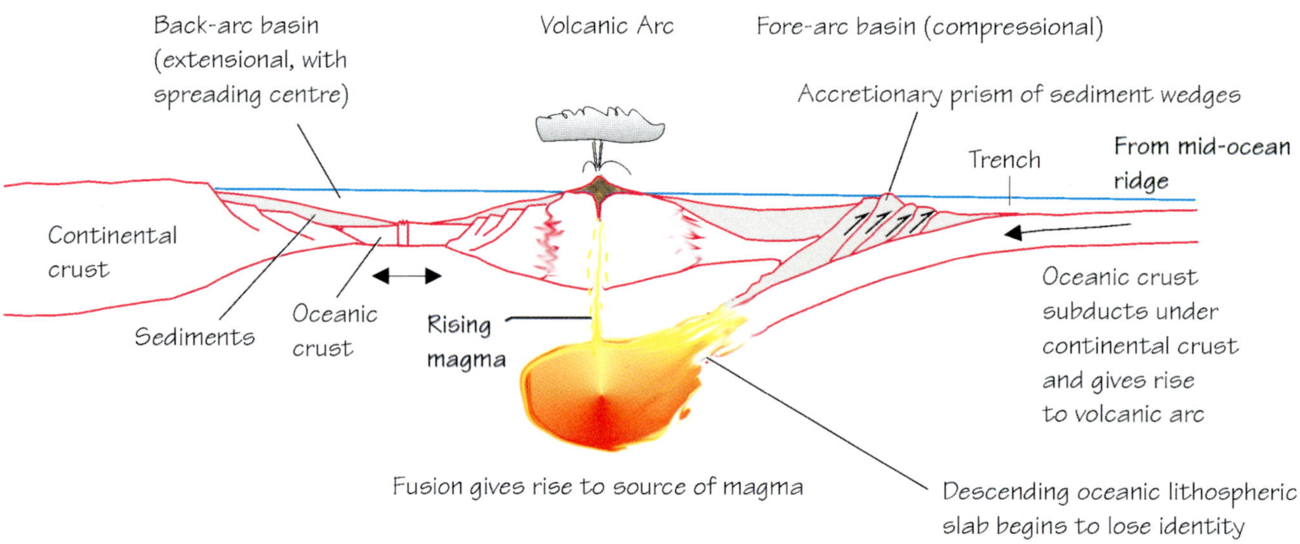

Fig. 3 Typical schematic cross-section through active continental margin (e.g. south-east Asia).

While the mantle is, to all intents and purposes, solid it is thought to 'flow' on a long timescale and thus drive the movements of the continental masses. However, no entirely convincing mechanism has yet been devised.

Fig.3 shows a schematic section through an active continental margin, with a variety of features. The oceanic plate on the right is sinking, or subducting, under the continental plate on the left.

Rocks, minerals and fossils

The earth is made up of rocks, which are composed of minerals and may contain fossils. All these are discussed at greater length below in the appropriate sections.

Principles of stratigraphy and dating

The sedimentary rocks that form the crust are conventionally thought of as a continuous pile of sediments, the 'stratigraphic column' (Fig.4). In practice, this is never found, and such an accumulation could not exist: it would be thicker than the thickest crust. This notional pile is split into convenient divisions as a result of the investigations of early geologists, with names given to the time divisions ('periods') and rock divisions ('systems'). These are subdivided in turn for convenient reference and classification. The diagram (Fig.4) shows such a

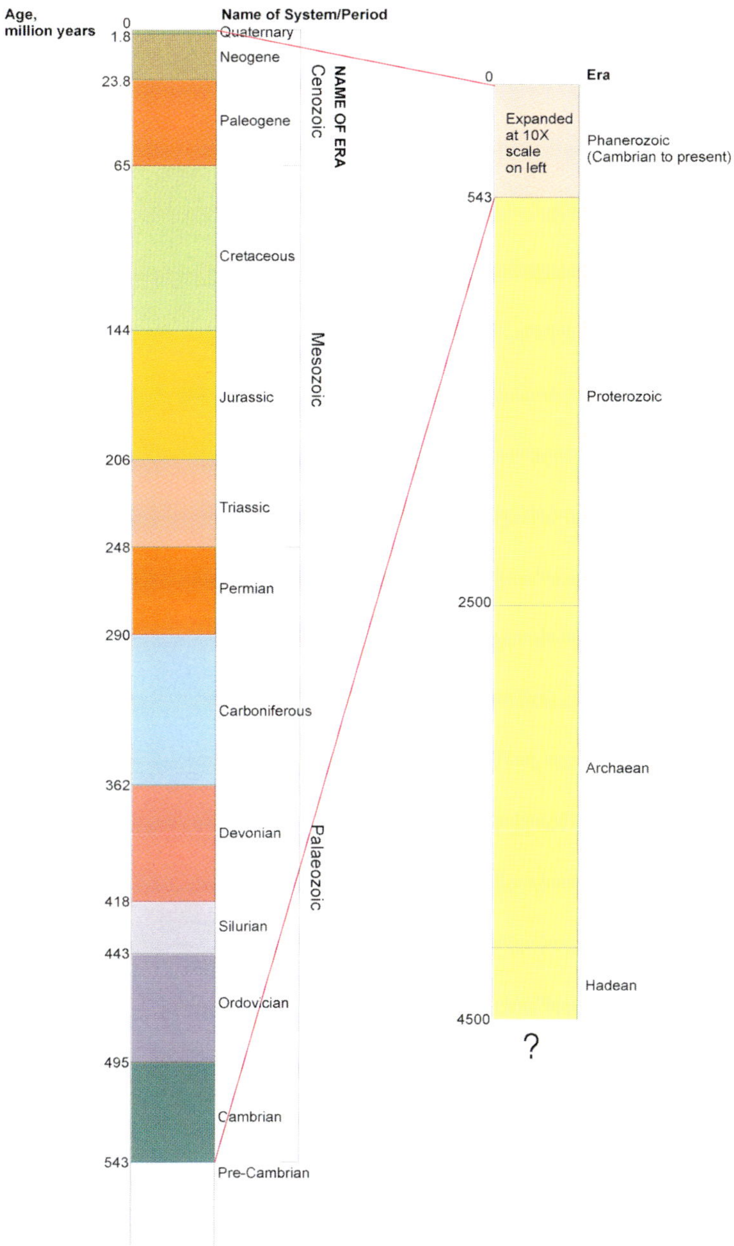

Fig.4 The Stratigraphical Column (the dates are subject to constant revision as geological knowledge increases). Constant reference to this diagram will increase your understanding of many sections of this book.

theoretical stratigraphic column. The size of each coloured block is proportional to the timespan of the named divisions, not to the thickness of strata deposited. It is convenient to treat the vast stretch of Pre-Cambrian time as one unit, since within the UK large outcrops of Pre-Cambrian rocks occur predominantly in northern and western Scotland (an outcrop is where a rock formation is exposed at the surface), well away from most urban areas.

Intrusions or lava flows of igneous rock were, until recently, dated only by reference to the sequence of sediments. Any sediment that has been intruded by a dyke, say, must pre-date it; while if the dyke is cut off at the base of other sediments, they must post-date it.

Pioneers in earth science included early Greek philosophers, the Italian Renaissance artist and thinker Leonardo de Vinci, and Georg Bauer, a mining expert of the 16th century.

Early geologists began to achieve a reasonable understanding of the earth's history and crustal processes by the early 1800s. They soon found that they could arrange the rocks into the correct time sequence. They achieved this by the principle of superposition (i.e. the sedimentary rocks on top must be younger than the rocks underneath) and by lateral correlation by means of rock type and fossil content (except for the Pre-Cambrian, where fossils are generally tiny and sparse). Similarly, they concluded that a rock that is intruded by an igneous rock must have been there earlier. William Smith, a canal surveyor and engineer, drew the first geological map of England in 1815, largely from his own knowledge.

Stratigraphy is the discipline of the study of rock successions. It attempts to discover the history of the earth by analysing the sequence of events that have resulted in sediments being deposited and fossil communities preserved. The terminology and classifications worked out by stratigraphers by about 1850 generally hold to this day, with minor modification. Within the UK, only relatively limited problems of dating or correlation now remain as far as most Cambrian and later rocks are concerned. In areas of older, metamorphosed rocks the techniques of stratigraphy are of less use and more arcane laboratory-based techniques are common tools.

Dating was, in the earliest days of geology, only relative, although it was soon realised that the famous date of 4004 BC for the earth's formation, calculated by Archbishop Ussher, was unrealistic. In the late 19th century the physicist Lord Kelvin estimated the age of the earth at only 20 million years; but his calculations were based on assumptions limited by the knowledge of his day. In the second half of the 20th century absolute (if rather approximate) dating became possible with the use of radioisotopes, which decay at a known rate. These radioisotopes are found principally in igneous rocks, although some can accumulate in sediments. This technique has lately become much more refined and accurate, and other dating techniques have been developed for correlative purposes. The range of tools at the laboratory technician's disposal now includes such arcane techniques as:

- Radiogenic isotopes, which decay at a known rate.
- Cyclostratigraphy, based on recognition of sedimentation cycles.
- Chemostratigraphy, which utilises chemical analyses to correlate rocks.
- Magnetostratigraphy, which depends on the changes over time of the earth's magnetism.
- Biostratigraphy, based on fossil correlation.
- Sequence stratigraphy, based on a theory of specific sediment blocks being deposited at different states of sea level.
- Event stratigraphy, which depends on recognisable events such as earthquakes and their effects.

In practice, you need not know the precise details of some of these techniques, which are generally impossible to apply in the field except in a few specific cases. Biostratigraphy, the use of fossils (see Chapter 7, Fossils) is the easiest for the layman to apply, crudely at least, as no special tools are required.

Table 1 lists the main types of time and time-rock divisions.

Table 1 Terminology for time and time-rock divisions

Time (Geochronologic) Division	Time-rock (Chronostratigraphic) Division: i.e. rocks deposited during given time division
Eon	Eonothem
Era (Superperiod)	Erathem (supersystem)
Period	System
Epoch	Series
Age	Stage
Chron	Chronozone (may be the same as a biozone)

In general, major time-division names and the names of the more important rock formations will be sufficient for the surveyor's purposes. To interpret geological maps it will be necessary to have some command of these names, although it is usually easy, with practice, to learn the sequence of formations in any limited geographical area. The stratigrapher will immediately grasp the significance of any well-drawn geological map presented to him.

The stratigraphy of the British Isles is discussed in more detail after rocks, minerals, and fossils have been covered.

Britain – a dynamic location

Britain came into existence in something resembling its current form only recently. Until about 10,000 years ago, or approximately at the end of the last major glaciation, it was still linked to France at the present Straits of Dover. As one goes back in time, Britain becomes less recognisable. Since Cambrian times alone, the rocks that now form the British Isles have been subjected to continental collisions and the formation of mountain ranges of Alpine dimensions. Volcanoes have erupted from time to time from island arcs like those of the present-day West Indies or the Philippines, leaving their eroded remnants for modern investigators to find.

The following sections will permit some basic understanding of the components that make up the earth. The stratigraphy of the British Isles is discussed in more detail in a later section.

Chapter 3 Minerals

Introduction

As with so much of geology, it will rarely be critical for the surveyor to possess sufficient expertise to be able to identify a particular mineral. Many rock types can be recognised with only a little knowledge, while others cannot be identified outside the laboratory. However, it is necessary to have some background knowledge to assist in rock identification and thus to assess mining issues, etc, with more confidence.

Mineralogists have classified minerals by many means (see below), as any given mineral has numerous properties that are either specific or fall within a defined range. Some minerals have very specific compositions and structures, while others have no fixed chemical formula or crystalline form. Even so, they may have a preferred style of formation, or habit. Minerals with very different physical properties (such as diamond and graphite) may be composed of the same elements (in this case, carbon).

Quartz (see Figs.5, 6 and 7), possibly the most common mineral, is colourless when pure, although minute impurities such as metal atoms can give attractively coloured varieties. Crystal lattice defects can also cause colour variation. Amethyst, rose quartz and the like are variants of quartz. Silica can also crystallise at low temperature to form what are termed 'cryptocrystalline' varieties, such as opal, jasper, onyx and flint. No crystals can be seen in these varieties, but a petrological microscope reveals the presence of crystal structure, i.e. an ordered arrangement of atoms.

Minerals may be of economic use (see Chapter 10, Economic Geology). While it is probably no longer likely that the surveyor will happen upon a valuable deposit that may make his client's fortune (or result in industrial activity that devalues his property), it is always possible. In Sutherland, alluvial gold has been found for years, but the source has not been traced. The ore-body would have to reach a concentration of about one part per million for mining to be considered viable. There is probably still economically workable gold in some unexplored part of the goldfield of North Wales, but much further exploration may be necessary to find it.

Economically valuable minerals may be an original constituent of the rock, as in some sedimentary iron ores; or may form later additions or insertions. Limestone, for instance, may be replaced by haematite (see below) and many minerals are found in veins that fill fault and joint cracks. In such cases, circulating water at various temperatures is thought to carry the appropriate material into place and to leach away any original material. Many readers will be familiar with the phenomenon of lime scaling, and to a certain extent (to simplify grossly) the process is similar.

The characters of minerals

Minerals may have characters depending upon:

- Visual appearance—colour, translucency, etc.
- Other human senses—taste, odour and touch.
- State of aggregation; crystal structure, hardness, fracture, etc.
- Specific gravity.
- Fusibility.
- Electromagnetic properties.

Visual appearance

A number of minerals have properties which, to the naked eye, allow them to be distinguished or at least suspected without further testing. This would include those which form well-defined crystals. However, colour is

Fig.5 This hexagonal crystal, with neat faceted end determined by atomic structure, is quartz, almost pure Silicon Dioxide (SiO2). The tiny bubbles (magnified) may contain gas or liquid from the time of formation that can be analysed and yield useful information. Diameter about 1 cm.

a property which may distinguish some minerals while other minerals may show a great variety of tints. Some minerals are clear, while others are almost completely opaque, even in a thin section prepared for the microscope. The quartz crystal (Fig.5) shows clarity and a regular crystal form which, together, are virtually diagnostic. The minerals may have a particular lustre, or way of reflecting light, which also may be diagnostic. The illustrations on the following pages show a wide variety of colours, forms and textures.

Other human senses

Occasionally, it may be possible to bring other senses into play. Some minerals have a distinctive taste, although it is not suggested that you try this, for obvious reasons. A crystal of rock salt will obviously taste of salt. When arsenic compounds are heated, they give off an odour reminiscent of garlic. A certain type of limestone which contains a high proportion of sulphides is known as 'stink stone' because of the smell of hydrogen sulphide when it is struck with a hammer and quickly thereafter sniffed. Some minerals may feel smooth or greasy to the finger.

State of aggregation

Except for Mercury and natural mineral oils, all minerals are found in a solid state. Under ideal conditions, most minerals tend to assume the particular geometrical forms of crystals, the identification of which can be an important tool. The flat (or occasionally curved) facets of the crystal are crystal faces. Compare Fig.5 with Fig.6. Although superficially the crystals seem different, measurement of the angles of the crystal faces would show that the same crystalline structure governs their arrangement.

Fig.6 Another type of quartz crystal

Many such crystals can be visually attractive, especially the larger forms. However, many crystals are so small, or interlock so intimately, that the crystal form is not necessarily very obvious. Occasionally, minerals are amorphous, that is they have a complete absence of crystalline structure. A wide variety of terms describes the particular variety in which minerals occur and these may not relate directly to their crystalline structure. This would include minerals which are, for instance nodular as in the flint of southeast England, or reniform (kidney-shaped), as in the variety of iron ore known as 'kidney iron ore' found in the Barrow area and elsewhere.

Minerals may also mimic the shapes of other minerals, where they occupy space which has been made available perhaps by the dissolution of the previous mineral.

Certain minerals, notably gold and copper, occur as the native metal itself.

Hardness can be reasonably diagnostic, since it varies greatly. There is a standard scale of hardness (the Mohs' scale) which is more or less arbitrary and uses relatively common minerals as reference points. For instance, the softest mineral, talc, can easily be scratched with the fingernail. The hardest mineral is diamond, which scratches all other minerals.

Certain minerals may be cut with a knife, beaten flat with a hammer, or bent. The way in which they fracture may be rough, smooth, or curved in a particular fashion. If the mineral has a crystal structure, it may break in a special way by 'cleaving' along definite planes which are determined by the crystal structure. These are not to be confused with crystal faces. Such minerals are said to possess cleavage, which may be variously qualified as, for instance, 'poor' or 'perfect'.

The crystalline structure of a mineral is usually intrinsic to its properties. As a mineral has a definite chemical composition, the way in which the constituent atoms are arranged gives rise to a more or less regular arrangement of bounding faces. Only under the most favourable conditions, however, does the full set of faces develop. The pattern of faces, and the angles between them, are directly affected by the crystal structure and can therefore be diagnostic.

Specific gravity

Specific gravity is easily assessed and can give a good guide to some minerals.

Heat properties (fusibility)

Certain minerals can be melted at the laboratory bench at more or less regular temperatures.

Electromagnetic properties

Magnetite is almost the only mineral affected by an ordinary bar magnet. However, numerous minerals are attracted in varying degrees by an electromagnet. Minerals containing iron are generally magnetic, but not always so. Minerals may have electrical properties: some are good conductors, while some are piezoelectric, that is they develop an electric charge when subjected to some form of stress. Finally, minerals containing elements of high atomic weight may be radioactive.

Some common or significant minerals

There are so many different minerals, each of which may occur in different 'habits' or typical forms, that only a thick book could do them justice. Most reputable bookshops of any size will at least carry some sort of pocket guide to minerals, but the following attempts to list some of the names that you may come across. If you study the figures carefully you should be able to recognise some of the material in old mine spoil heaps, as well as a few of the more common minerals that may be found in the coarser-grained rocks or in veins.

Many minerals, especially economically important ones, have a number of names and alternatives are given in italics where noteworthy.

Silicate minerals

All silicate minerals include silica, but otherwise their formulae can be very complex, usually including sodium, potassium, magnesium, calcium and aluminium in various proportions. They tend to form in groups of highly variable composition but with commonalities of structure or part of the chemical formula. Most of them are rock-forming, that is they contribute to the general mass of an igneous or metamorphic rock.

Fig.7 Typical appearance of milky quartz as fracture fillings in contorted and indurated Ordovician mudstones, north Pembrokeshire. Such veins are commonplace in contorted older rocks.

Quartz (Figs. 5, 6 and 7)

If you have a sandy beach nearby, go and pick up a handful of sand. The chance is 99% that it will consist almost entirely of grains of quartz. Quartz is the most resistant common mineral and crystallises in igneous (mostly acid, or granitic) or metamorphic rocks. It often forms veins (Fig.7), which may be very thin or of the order of a metre thick. These veins may be mineralised (in the sense of minerals of economic value). Although they may be separated from their original rock by weathering and erosion, individual grains persist because of their hardness and can be incorporated in sandstone and re-eroded, perhaps through several cycles.

Most mature sandstones consist largely of quartz; the whiter they are, the more likely they are to be a hard, resistant silica-cemented quartz sandstone, i.e. nearly pure silica. Varieties of quartz include amethyst and other semi-precious stones. The colour is usually due to impurities or inclusions.

Felspar (=Feldspar) (Fig.8)

Felspar also crystallises in igneous and metamorphic rocks. There are many varieties, mostly pale-coloured, and often pinkish. The fundamental grouping is into Orthoclase (or sodic/potassic felspar) and Plagioclase (calcic felspar), found respectively in more acid and more basic rocks. Some varieties, like moonstone, can form attractive crystals or gemstones. The ornamental 'Norway Porphyry' or Larvikite consists largely of crystals of a variety of feldspar called labradorite. This produces a beautiful play of interference colours due to its internal structure.

Fig.8 Orthoclase felspar.

Mica

Mica is found in igneous and metamorphic rocks of many types, and is loosely divided into the Muscovite group, which is paler in colour, and the biotite group, which is darker. It has a layered atomic structure, and cleaves perfectly into very thin sheets. This property led to its use in observation peepholes in boilers, electrical insulation, etc. When eroded and redeposited, the small flakes form reflective, sparkly dustings on bedding planes. In many metamorphic rocks, crystals form at right angles to the main pressure direction and give an obvious laminated structure.

Asbestos minerals

Asbestos is a commercial term for a variety of silicate minerals that have long fibres: these include

- chrysotile, a fibrous serpentine (serpentine is magnesium silicate formed by alteration of certain basic igneous rocks);
- actinolite and tremolite, both amphiboles (see below);
- amosite and anthophyllite, another amphibole group;
- crocidolite, a variety of riebeckite, yet another amphibole. This is the notorious 'blue asbestos'.

Amphiboles (Fig.9)

This large group includes such names as hornblende, the principal member, as well as the varieties named above as asbestos. The minerals are usually dark-coloured; they are related structurally but have widely varying formulae. They form in both igneous and metamorphic rock.

Fig.9 A mass of greenish interlocking hornblende crystals.

Pyroxenes (Fig.10)

Another large and variable group of mostly dark minerals with related structures, forming igneous and metamorphic rocks; augite is the principal member. The dark crystals in gabbro are usually augite or a related pyroxene.

Fig.10 (left) Individual crystal (rather battered) of augite. The edges of the crystal faces are just discernible. Fig.11 (right) Garnet of Almandine variety. About 2 cm across.

Olivine

A group of magnesium iron silicates mostly found in basic to ultrabasic rocks (for instance gabbro); they are mostly greenish and massive but can be of gem quality. One variety is formed when dolomitic limestone is metamorphosed and subsequently alters to give the green streaks in Connemara Marble.

Garnet (Fig.11)

Garnet is usually an accessory mineral in igneous and metamorphic rocks, but is occasionally a major part of some rock types. Garnets form a group of reddish or brownish minerals that can be cut and polished as gemstones.

Clay minerals

This is an extremely variable group of hydrous aluminium silicates. They result from the alteration or weathering of rocks, mostly of igneous origin. The weathered material can remain in place as a residue, or can be transported and redeposited. The individual minerals are best distinguished by X-ray diffraction as they are so fine-grained.

Clays become plastic when wet and can then be moulded and made into pottery, bricks, etc, by firing. They may then also be subject to flow or shearing processes that give rise to landslides (q.v.).

Many clay minerals have the property of adsorbing water onto their crystal lattices, which can increase their volume substantially. These are the notorious shrinkable clays. When desiccated by persistent dry weather and/or vegetation moisture demand, they shrink alarmingly. Montmorillonite, a member of the smectite group of clays, is most affected.

Non-silicates

Calcite (Figs.12, 13 and 14)

Calcite, calcium carbonate, is the most common carbonate mineral and is the principal constituent of most limestones and marbles. Many shellfish have calcite shells. It may also cement some sandstones, and occurs widely as veins. In practice it usually includes small amounts of other elements that can fit in the crystal lattice, such as magnesium. It occurs as biogenic (often massive or earthy) forms in limestone, where it forms the mass of the rock. It can also form well-defined crystals and may be the principal gangue mineral (i.e. the valueless matrix) in ore veins.

Cleaved fragments usually form well-shaped rhombs, i.e. with roughly diamond-shaped facets, and have been much used as 'spar dash' on rendered finishes.

Fig.12 Calcite: weathered cleavage rhombs (larger about 1.5 cm wide) picked off a mine dump.

Calcite is the main constituent of stalactites and

Fig.13 (top left) Calcite vein in limestone (note microfaulting and slight calcite mineralisation of fault).

Fig.14 (top right) "Beef" calcite (pale band across centre) in Lower Lias shales, Dorset.

Fig.15 (bottom left) Aragonite. About 5 cm across.

Fig.16 (bottom right) Dolomite.

stalagmites in caves (as well as a major part of flowstone formed by leaching of calcium carbonate from concrete). Tufa, a calcareous material formed by deposition from hot springs, can sometimes form thick beds (travertine) which are used as a floor or wall finish within buildings.

The figures show only a few of the ways in which calcite can appear.

Aragonite (Fig.15)

Aragonite is also calcium carbonate but is not an important rock-former. Its crystalline structure is different from that of calcite. It is less stable than calcite and readily converts to it. However, much biogenic limestone begins as aragonite within shells or coral.

Dolomite (Fig.16)

Dolomite is calcium magnesium carbonate and is usually formed by the alteration of limestone by replacement of calcite. Dolomite is denser and such dolomitisation can give rise to cracks in which useful minerals can be deposited. Dolomite is commercial useful, for instance in the manufacture of furnace linings.

Barite *Barytes, Heavy Spar* (Fig.17)

Barite is termed heavy spar because it is unusually heavy for a light-coloured mineral; it is halfway between most minerals and iron ore in density, so that the difference is obvious when a piece is held in the hand. It is barium sulphate, usually with some strontium and calcium. It can form tabular crystals but often occurs massive (i.e. a more or less anonymous lump), granular, etc. It is usually white or colourless but can be faintly tinted yellow, red or brown. It occurs as a common vein material in lead and zinc veins. It is commercially important and used in paint, dressing fabric and paper, and as drilling mud (the principal use). Extraction has occurred in many parts

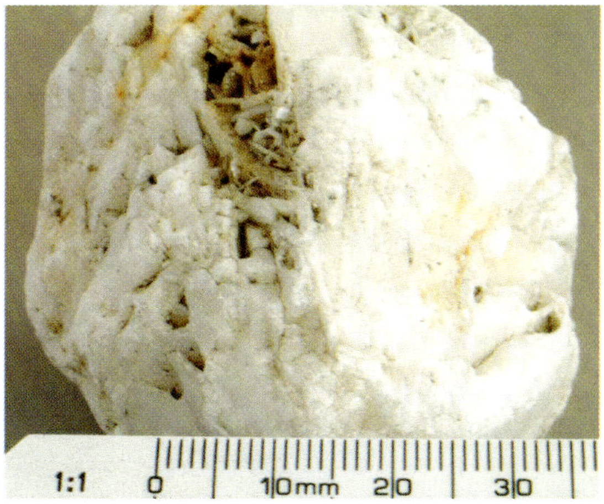

Fig.17 Barite (picked off mine dump).

of upland Britain; between 1940 and 1946 over a third of UK production came from veins at Muirshiel in the Renfrewshire hills and Gasswater near Muirkirk.

Gypsum

Gypsum is hydrous calcium sulphate that is formed as a saline residue or evaporite (see below), as a by-product of dolomitisation limestone, or within clays to form the well-defined crystals (*selenite*) illustrated by Fig.18. Within clays, pyrite decomposes and the resulting sulphuric acid attacks any calcite present (such as shells) and converts it to calcium sulphate. Gypsum is industrially important (e.g. for the production of plasterboard).

Anhydrite

Anhydrite is anhydrous (hence the name) calcium sulphate. It may form well-defined crystals but usually forms various fibrous or massive varieties. It is usually a saline residue associated with gypsum and rock salt, and is extracted for chemical and industrial purposes.

Fig.18 Gypsum (hydrated calcium sulphate) crystals, variety selenite. Brown staining from location in London Clay. These crystals can be dissolved in hot water. Each about 1.5 cm wide.

Halite

Halite (rock salt) is sodium chloride, common salt, but in nature usually contains some calcium and magnesium chlorides, as well as calcium sulphate. It occurs as an evaporate following the drying-up of enclosed or partly-enclosed seawater bodies. It commonly forms cubic crystals, sometimes with hollowed, stepped, faces (hopper crystals). Salt occurs in sufficiently large bodies to form rock that can flow under relatively low stresses and (as they are less dense) sometimes burst through overlying strata (forming a 'diapir'). It is, naturally, economically useful and can be associated with formation of oil traps in oilfields. Salt can be mined, as on the continent, or extracted by brining - dissolving the salt in pumped water and then re-evaporating it at the surface. Rock salt extraction in Cheshire has given rise to considerable subsidence, commonly with the formation of small lakes. Droitwich in Worcestershire was an early salt extraction centre, known to the Romans.

Apatite

This mineral is a complex fluoro– and/or chlorophosphate of calcium and is found widely in the British Isles, but in small amounts, in igneous and metamorphic rocks and limestones. Human bones and (especially) teeth contain organically-formed versions of this mineral.

Fluorite (= Fluorspar), *Blue John*, *Derbyshire Spar* (Fig.19)

Calcium fluoride forms cubic crystals, usually in mineral veins, some well known as in Derbyshire where banded varieties are used to make jewellery. It also cements some sandstones and is industrially significant (e.g. optical lens manufacture). Extraction, partly under urban areas, continues today.

Fig.19 Interlocking crystals of Fluorite. Scale in mm.

Pyrite, *Fool's Gold*; Chalcopyrite

These minerals are respectively iron sulphide and copper iron sulphide.

Fig.20 Interlocking cubic crystals of pyrite. Also crystallises as the distinctive pyritohedron, with five-sided crystal faces.

Pyrite (Fig.20) is known as 'fool's gold' as it has deceived many hopeful innocents in the past by its brassy lustre. It is also known as mundic, the material that became incorporated in so many concrete blocks in Devon and Cornwall, which subsequently failed as the pyrite oxidised. Pyrite is commonly found in ore veins. Although the mineral contains much iron, it is useless as iron ore because of the sulphur content. Occasionally it contains payable quantities of gold. It is present in many shales and clays, and gives rise to sulphuric acid activity and reddish mine discharges (see below). Some fossils are preserved in fine detail by pyrite.

Chalcopyrite is also a brassy-yellow colour, but has other distinctive properties. It is the main source of commercial copper and has been worked patchily and generally on a small scale in many of Britain's highland regions, with occurrences also in the Triassic hills of Cheshire. The heart of the Great Orme, the promontory next to Llandudno, has been mined since the Bronze Age. Devon and Cornwall produced many rich, and quite large, mines; a few of these worked until quite recently but became uneconomic although ore reserves remain. Other small deposits occurred in Shropshire, North Wales, and the Ochil and Renfrewshire Hills in Scotland's Midland Valley; and in parts of the Southern Uplands. Avoca, Co Wicklow, was one of the largest mines in Ireland and produced Chalcopyrite.

Malachite

This distinctive greenish copper carbonate can be found in a banded variety. It is copper-rich and valuable, but usually occurred near to the surface above chalcopyrite, so that it has been worked out. Greenish streaks on rocks in old mining areas are often an indicator of malachite and by extension of other copper ores.

Haematite (Hematite) (Fig.21)

This is a form of iron oxide, with various forms including so-called 'kidney' ore as well as massive red varieties. It is rarely found in the UK in significant quantities but has been extracted in the past in South Wales, the Forest of Dean, and near Ulverston. In all of these places it has replaced limestone. In Scotland, veins have been worked near Haddington and near Sorn in Ayrshire. Red streaking in exposed rocks may indicate the presence of haematite.

Magnetic iron ore, magnetite, is not significant in the British Isles.

Fig.21 Haematite.

Sphalerite *Blende, Zinc Blende, Black Jack* (Fig.22)

Sphalerite is zinc sulphide, but usually with a little iron and cadmium. It is the common ore of zinc and is usually found associated with galena, and often as a limestone replacement. It has been widely mined on a small scale in this country, including in Cornwall, Cardiganshire, Shropshire, Derbyshire and Cumbria.

Galena *Lead Glance, Blue Lead* (Fig.23)

Galena is lead sulphide, but there is usually some silver replacing the lead, sometimes to a sufficient degree to make recovery worthwhile (argentiferous galena). The mineral is the principal source of lead. It occurs, often associated with Sphalerite, in vein or limestone replacement deposits. Barite is often a significant gangue mineral. Mining in Britain has occurred in most upland counties, but most notably in various parts of Wales, Somerset/Avon, Shropshire, Derbyshire, Yorkshire, and Lanarkshire. In the Leadhills area, considerable reserves are thought to remain. Tyndrum in the Highlands has yielded considerable quantities. Low-grade disseminated strata-bound mineralisation has been noted in the Lower Carboniferous of Scotland's Midland Valley. Galena can form well-defined silver-grey cubic crystals and also has perfect cubic cleavage, such that when struck it often disintegrates into small cubes.

Cassiterite *Tinstone*

In Britain, Cornwall and Devon were for many years the focus of intensive mining activity based on cassiterite, tin oxide, usually found as black or brown crystals in quartz veins associated with granitic rocks. It has also been extracted from alluvial deposits (placer mining). Tin mining in Cornwall experienced a late resurgence after 1970, at which time several mines were developed, together with placer extraction and waste reprocessing. However, as the artificially inflated price of tin collapsed again, work ceased. The last mine, South Crofty, closed in 1998. Its workings extend to a depth of 1 km below Redruth and Camborne.

Fig.22 Sphalerite.

Fig.23 Galena with distorted cubic cleavage.

Chapter4 Rock types

Igneous

Magma

Rock within the earth's crust can become molten at great depth, at temperatures of about 800°C. As processes such as plate movement occur, some sections of the lower crust are preferentially heated. As a result, they become less dense and can rise nearer to the surface.

Such molten rock is termed magma. The composition of the magma depends on the original material, the material through which it passes as it rises, and other physical factors. Magma that reaches the surface may erupt at a volcano (see specific section on volcanoes) to produce volcanic rock. The rock may erupt as fluid, termed lava. If the magma contains dissolved gases, these can expand explosively as they experience pressure reduction at the volcano's mouth. This explosive force can blast the magma into fragments (which can be quite tiny), termed volcanic ash (even though combustion as such does not take place). This ash falls back to earth to form pyroclastic deposits.

Magma that remains below the surface and cools there forms an intrusion. Fig.24 shows in diagrammatic form the principal types. Intrusive rocks are termed plutonic but this term has become restricted to larger masses (10 metres upwards, or thereabouts). Lesser masses (which can be as thin as a centimetre or so) are termed hypabyssal. The term minor intrusion is also found in general use.

Very large intrusions (which normally consist of a series of amalgamated smaller intrusions) are termed batholiths; there is a bewildering number of terms for lesser intrusions, but the principal ones are dykes, which are subvertical and sheet-like; and sills, which are generally concordant with sedimentary bedding and are also sheet-like.

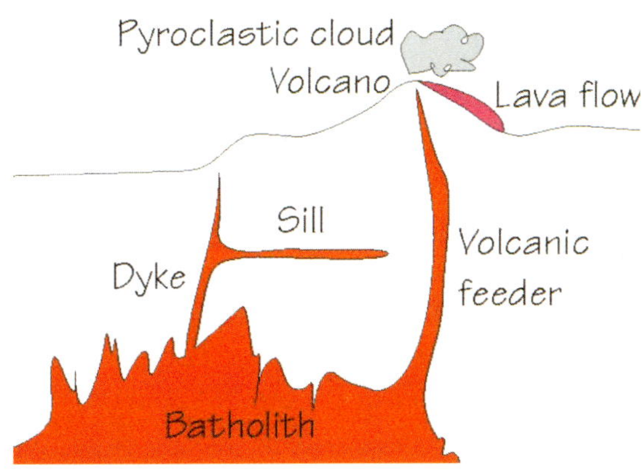

Fig.24 Highly simplified sketch defining major occurrence modes of igneous rocks. (NB Dykes and sills are essentially planar bodies.)

Although intrusions form at depth, subsequent changes in the crust have made older intrusions accessible and available for scientific study.

Deeper rocks are exposed as the earth's surface is eroded by rain, frost, thermal processes, chemical processes, glaciers, biological activity and so on, thus revealing intrusions such as the granite of Dartmoor. Earth movements, like mountain-building, lift such rocks to the height at which erosion can occur.

Crystallisation

As the rocks cool, they crystallise, forming a mass of interlocking crystals dominated by minerals based on silicates (see Chapter 3, Minerals). These include strong tetrahedral groups of silicon oxides, SiO_4. As the crystals grow by migration of molecules through the molten rock, they interfere with each other. Those crystals that form first (i.e. have the lowest crystallisation temperature) have the best-formed crystal shapes; those that crystallise later are forced to occupy the remaining 'spaces', and do not have a well-defined crystal form, although they do have an ordered crystal lattice structure of molecules. The interlocking nature of the crystals produces great toughness and resistance to erosion.

As a rule, rocks formed at greater depth have cooled slowly and have coarser crystals, which may be exceptionally well-formed. This is because molecules have more time to migrate and arrange themselves in orderly fashion. On the other hand, some lavas cool very rapidly and are so fine-grained (aphanitic) that the crystals are too small to be seen by the naked eye; in some cases, there is no time for crystals to form and the rock is glassy.

In general, basic rocks tend to be rather finer-grained than their acid depth equivalents.

Where magma begins to cool at depth so that crystals begin to form, but is then forced by pressure to a cooler location, this results in obvious crystals, or phenocrysts, in a finer groundmass.

Many of the silicate minerals are inherently highly resistant to weathering and erosion. Quartz (near-pure silica, SiO_2) and some other minerals (notably felspars) are resistant even when weathered out of the rock mass, and are redeposited as sediments.

In general, rocks of a given chemical composition can be found at all grain sizes - but not necessarily so. Crystals of different compositions form at different temperatures, so that in any rock of the same overall composition, usually the first-formed, and therefore the best-formed, crystals will be the same mineral.

Classification

In the early days of igneous petrology, any apparently distinctive type of rock newly discovered would be given a new name, so that a bewildering variety arose. The tendency of some researchers now is to rationalise names. For instance Essexite (named not after Essex, England, but after a county in Massachusetts, USA) is a particular variety of gabbro which is defined by reference to the additional minerals that it contains. It would be much more explanatory to call it a nepheline-analcite-olivine-gabbro, although that is admittedly unwieldy. However, many obscure names still persist and may appear on geological maps.

The presence and percentage of quartz and of the different types of felspar are generally important, with subdivision according to the presence of subordinate minerals.

Coarsely crystalline rocks can often be identified approximately in the 'hand specimen' because the mineral grains are so large that they can be seen individually. Most coarse igneous rocks are granitic in the loosest sense. Rocks which contain large crystals, often whitish or pinkish-red, with clearly-defined edges, are usually granite (compare Figs.25 and 26). The urban surveyor will be familiar with the use of polished granite slabs in building facades. Study of such slabs even with the naked eye often allows three or four individual minerals to be distinguished if the crystals are large enough. Kerbstones and paving setts may also be of granite and study of even these unpromising items can be informative, as very coarsely crystalline rocks are often used.

Among the coarser-grained igneous rocks, granite, and rocks closely allied to it, are predominant in this country.

Finer-grained rocks may be virtually impossible to identify. Most igneous rock masses will occur as intrusions, but the lavas and pyroclastic materials will be found interbedded with sediments.
In such case, they may not immediately be separable to the eye (even, at first sight, to some professional geologists).

Other igneous types occur more rarely in these islands. Dolerite (see Fig. 27) is one of the more common: it is often greenish black, rather than merely almost black (partly due to subsequent

Fig.25 (top) In this polished specimen of granite, the view is dominated by large pink feldspar crystals, which are intergrown with quartz. The glassy crystals (notably top right) are quartz; the dark crystals are mica and other minerals. Fig.26 (below) Dartmoor granite: hand specimen. Note the very large felspar crystal upper left.

Fig.27 (right) Dolerite: much used for roadstone. Note brownish weathered surfaces. Titterstone Clee, Shropshire.

Fig.28 (right centre) Arthur's Seat, Edinburgh: the peak is the eroded stump of a volcanic vent pipe. Salisbury crags, in the mid-ground, are formed by a dolerite sill. Both are of Carboniferous age.

Fig.29 (far right, top) High Force, where the Tees falls over the edge of a branch of the Whin Sill.

Fig.30 (bottom left) Gabbro, with distinct greenish tint.

Fig.31 (far right bottom) Andesite, with porphyritic crystals.

alteration), and may have crystals that can just be picked out by the eye. It is often found as roadstone (you may find stockpiles at highway depots) or paving setts, as it is hard-wearing due to its particular crystalline arrangement. Dolerite often occurs as sills (Figs.24 and 28) and dykes, and where these are thick enough it tends to form columnar jointing at right angles to the margins of the intrusion. These can be seen at the Giant's Causeway, in Ulster, and at the famous Fingal's Cave on the Scottish island of Staffa. Fig.29 shows a weakly developed columnar dolerite. Gabbro, the coarse-grained equivalent (Fig. 30) is rarely seen.

Andesite can often be guessed at in the field since although it is predominantly dark and fine-grained it often carries small but distinct pale felspar phenocrysts (Fig.31), of a variety termed plagioclase.

Many igneous rocks, however, have unidentifiably small crystals and can only be identified precisely in the laboratory.

Volcanoes can erupt underwater. One of the resultant types of lava (found generally only in western and northern Britain) is pillow lava (Fig.32), where the lava (usually a type of basalt) naturally forms large blobs under the water that are about the size and shape of a pillow.

Fig.32 Pillow lava: the shapes are quite distinct on the right. Strumble Head, Pembs.

In general the igneous rocks are hard, tough or splintery, and form poor aquifers. They are usually excellent foundations for buildings, although they may be weakened locally by features such as faults or joints, or by chemical weathering. For instance, the granites of Cornwall have been chemically 'rotted' locally to a mass of china clay (predominantly the clay mineral kaolinite, which has been extracted as a material valuable in its own right). Granitic rocks also emit radon gas (see box). Many igneous rock bodies are associated with mineralisation. This can indicate a risk of mining voids, landfill and various types of contamination such as heavy metal or arsenic.

Igneous rocks are not common outside Wales, Devon/Cornwall, the Lake District, northern and western Ireland, and Scotland. However there are notable areas of igneous material locally, such as the dolerite Whin Sill in Northumberland and the Caldecote Volcanics near Nuneaton. The Whin Sill, in particular, outcrops in places over a large area of northern England, from Cross Fell to Bamburgh, and underlies at depth much of the intervening ground. Some of its outcrop forms a dramatic scarp topped by Hadrian's Wall. In places it is paralleled by a lesser sill, the Little Whin Sill. With the exception of the areas of Devon and Cornwall, in England and Wales large igneous masses are very rare and many surveyors will probably never have to take igneous rocks into account. Scotland, Ulster and Leinster feature much more extensive areas.

Occasionally, very small, isolated, bodies of igneous rock are to be found in the middle of otherwise entirely sedimentary areas. Some of these probably remain to be mapped. They are often thought to be the feeders of long-vanished volcanoes.

Table 2 (opposite) shows a list of the more common igneous rock names by composition (based on silica content) and grain size. There are numerous finer divisions that are of interest only to the specialist, and there are other ways of classifying, but most geologists will be familiar with the table. The terms 'acid' and 'basic' are archaic but still much used as convenient shorthand.

Radon gas

'The ever-increasing evidence is that long-term exposure to elevated radon levels is linked to an increased risk of lung cancer and this is accepted by authorities and experts throughout the world.' Radon Atlas (cited in full below).

The reader should be aware of the possible risk to health posed by radon, a dense odourless gas which issues naturally from the ground in very small quantities.

Radon is generated by the decay processes of uranium-238, an element that is incorporated in certain minerals that are found mostly in igneous rocks but also in some clays. It may also be present in groundwater. At several stages during the decay sequence, potentially damaging subatomic particles or radiation are generated. As the gaseous components of the sequence are breathed in, they may decay to daughter metals, which lodge in the lungs and emit damaging alpha-particles. It is thought that the risk of developing cancer, primarily of the lung, is thus increased by exposure to the gas, although different authorities dispute the significance. As research has proceeded, the areas of the country subject to significant levels of the gas are becoming better defined. Building Regulations now require, as building surveyors will know, suitable forms of construction in Radon-affected areas.

The latest *Radon Atlas of England and Wales* (National Radiological Protection Board, 2002) shows high concentrations, as would be expected, in granite and other igneous areas such as the Cheviot granite, and also a distinct correlation with limestone areas, including some areas of Chalk. This is thought to be due to the solubility of the gas in groundwater. There are also distinct correlations with other geological features, but these are not necessarily so obvious. For instance, a line of high readings along the Bala fault in north Wales is presumed to be associated with the presence of deep fracture systems that allow the gas to rise to the surface.

Volcanoes

A volcano occurs where magma (molten rock) approaches the surface of the earth's crust and, ultimately, perforates it (an eruption). Most volcanoes occur in areas where the crust is being particularly stressed or heated -mostly at plate boundaries, rift valleys and major fault lines. There are various types of eruption, which vary in levels of destructiveness according to the material erupted, etc.

The erupted material may be lava, pyroclastic material (often termed 'ash', although it is just pre-existing material blasted apart by the eruption) or pyroclastic flows (*nuée ardente* = French, roughly, fiery cloud) in which particulate material flows

Table 2

Simple Classification of Igneous Rocks

Mode of occurrence	< More silica ("acid"), lighter colour		Less silica ("basic"), darker colour >		Grain size
Lava	**Rhyolite**	**Dacite**	**Andesite**	**Basalt**	**Fine-grained**
Dykes, Sills	**Microgranite**	**Microgranodiorite**	**Microdiorite**	**Dolerite**	**Medium-grained**
Larger Masses	**Granite**	**Granodiorite**	**Diorite**	**Gabbro**	**Coarse-grained**

suspended in hot gas. An example of the latter occurred during the Mt St Helens eruption. Destruction is devastating. Further damage may be caused by lahars, or volcanic mudflows, which mobilise fresh pyroclastic material. The form of the volcano is dictated by the type of eruption, which in turn is dictated by the material erupted. Basalt, which is erupted as a readily flowing liquid, produces volcanoes that, like those of Hawaii, have very gentle slopes, while the classic volcano shape beloved of cartoonists, such as Japan's Mt Fuji, is the product of much more viscous andesitic lava and pyroclastic material.

Volcanic rock that is not close to its centre of eruption may simply form a thin layer within the 'pile' of sediments in a region and this can be very useful for radiometric dating purposes, as the 'absolute' dates obtained in this way can be matched to relative dating derived from fossils or other techniques.

Some 60 million years ago massive volcanic outpourings took place in the present Ulster and Hebridean areas as Britain rifted away from North America and the North Atlantic extended. However, the likelihood of volcanic action in the British Isles at present is extremely small, probably vanishingly so. To all intents and purposes it can be discounted. The nearest recently active volcanoes are in central France and western Germany. However, the extent of very old volcanic rocks in the British Isles is considerable, mainly in the older, highland areas of western Britain.

As an aside, terms such as 'extinct' and 'dormant' should be used with care. As a generalisation, a dormant volcano may erupt again; an extinct one certainly will not.

In the extreme, a volcano literally blows itself to bits, as did Krakatoa (Krakatau) in 1883. This leaves a huge crater-like hole, called a caldera, which may be filled by sea or lake water. Santorini (Thera) in the Aegean is a caldera, the remains of a massive volcano that destroyed itself in about 1500 BC, showering the Aegean with pumice. Three islands form the remains of the rim, while a small, mildly active central island is the puny successor to the original volcano. There are, however, no recent examples in the British Isles.

Sedimentary rocks and sedimentology

Sedimentary rocks—introduction

Sedimentary rocks include a number of materials that are not strictly sediment, in the commonly understood meaning of waterborne particles, but the term is used by geologists, nevertheless, as a convenience.

Sedimentary rocks may perhaps be defined as rocks that are neither igneous nor metamorphic. They are stratified, i.e. laid down as layers with thicknesses that may vary but in general are about the same over a distance of (say) a kilometre. This means that one type of rock can be traced from one area to another, sometimes for hundreds of kilometres, although that is unusual in the British Isles.

The subdivision of sedimentary rocks is complex.

The different types of sediments have highly variable gross qualities that determine their performance under load, their permeability, etc. Two examples of the complexity of sedimentary lithological classification, for sandstones and limestones respectively, are given here (Table 3 and Fig.33, see overleaf). These need not be understood, but must be taken as an indicator that the rock formation referred to on a map, etc, need not have the properties assumed from possibly loose terminology. Before making hasty assumptions from a map, it is wise to make sure just what the map key really signifies.

Table 3

Folk's classification of carbonate rocks by composition		
Principal grains in limestone	**Cemented by: sparite**	**Cemented by: micrite**
Skeletal grains (bioclasts)	**Biosparite**	**Biomicrite**
Oöids	**Oösparite**	**Oömicrite**
Peloids	**Pelsparite**	**Pelmicrite**
Intraclasts	**Intrasparite**	**Intramicrite**
Limestone formed in situ	**Biolithite**	**Fenestral limestone—dismicrite**

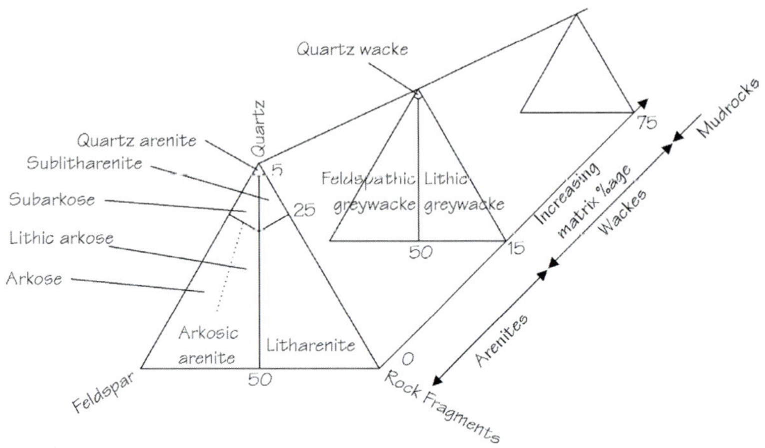

Fig.33 Classification of sandstones according to Pettijohn et al. The numbers refer to percentage compositions.

Sedimentary rocks are (notwithstanding the comment above) mainly deposited as sediment, usually under-water but sometimes on land or in lakes. They are divisible into:

- Siliciclastic rocks, formed by deposition of grains of sand, silt or clay derived by erosion of existing rocks.
- Carbonate rocks, formed by chemical and/or (more usually) biological processes.
  (Between them, these first two make up the bulk of sedimentary rocks.)
- Evaporites, formed by evaporation of sea or lake water.
- Sedimentary iron deposits and phosphate deposits.
- Coal, oil shale and petroleum.
- Cherts and siliceous sediments.
- Volcaniclastic sediments.

Sedimentary rocks defy a single, simple classification. Siliciclastic rocks range from the softest clay (Fig.34) to the hardest sandstone (Fig.35). Quartz arenite is as hard as the hardest igneous rock. Limestones include the massive, hard, limestones of the Mendip Hills or the Peak District, and soft, shelly oölites from the Cotswolds. Evaporites include the salt beds of Cheshire, never visible at the surface but a major cause of subsidence. Subsidence caused by coal extraction in many parts of Britain affects buildings directly and indirectly (by destabilising slopes and encouraging landslides). The particular type of chert termed flint, which is found within the chalk formations (another limestone) of southern and eastern England, is often used as a building stone. These various rock types are discussed in more detail below.

Fig.34 London Clay: oxidised specimen about 120 mm across. Freshly exposed clay is bluish or greyish.

The schematic landscape/cross section (Fig.36) shows how rocks may be arranged, based on their original depositional setting: different rocks can be deposited simultaneously within

a relatively short distance of each other. Conversely, some rock formations can be persistent over wide areas (the Chalk is a good example). The diagram shows, very crudely, how different sedimentary rock masses may pass into each other by interfingering. Gradational boundaries, in both vertical and horizontal directions, are also possible. In some cases, interfingering may be on such a fine scale that foundation conditions may change significantly several times in the area of a single building site.

It is important to bear in mind that lithology is not a respecter of time. While sand may be deposited as, say, a river mouth bar at one time, simultaneously mud may be deposited immediately adjacent. As the river deposits more sediment and the coastline moves outwards, the mouth bar will be further out, so that the eventual body of sandstone uncovered by geologists will be of different age at different locations; it is said to be diachronous or diachronic.

Sedimentology

Sedimentologists study the nature and origins of sediments. Study of grain sizes, types of bedding, rock distribution patterns and other features allows them to interpret past environments, and to predict underground variations (such as locations where oil drilling might be profitable).

Fig.35 (above) Resistant sandstones form this Derbyshire 'edge'.
Fig.36 (below) Common depositional environments (greatly simplified).

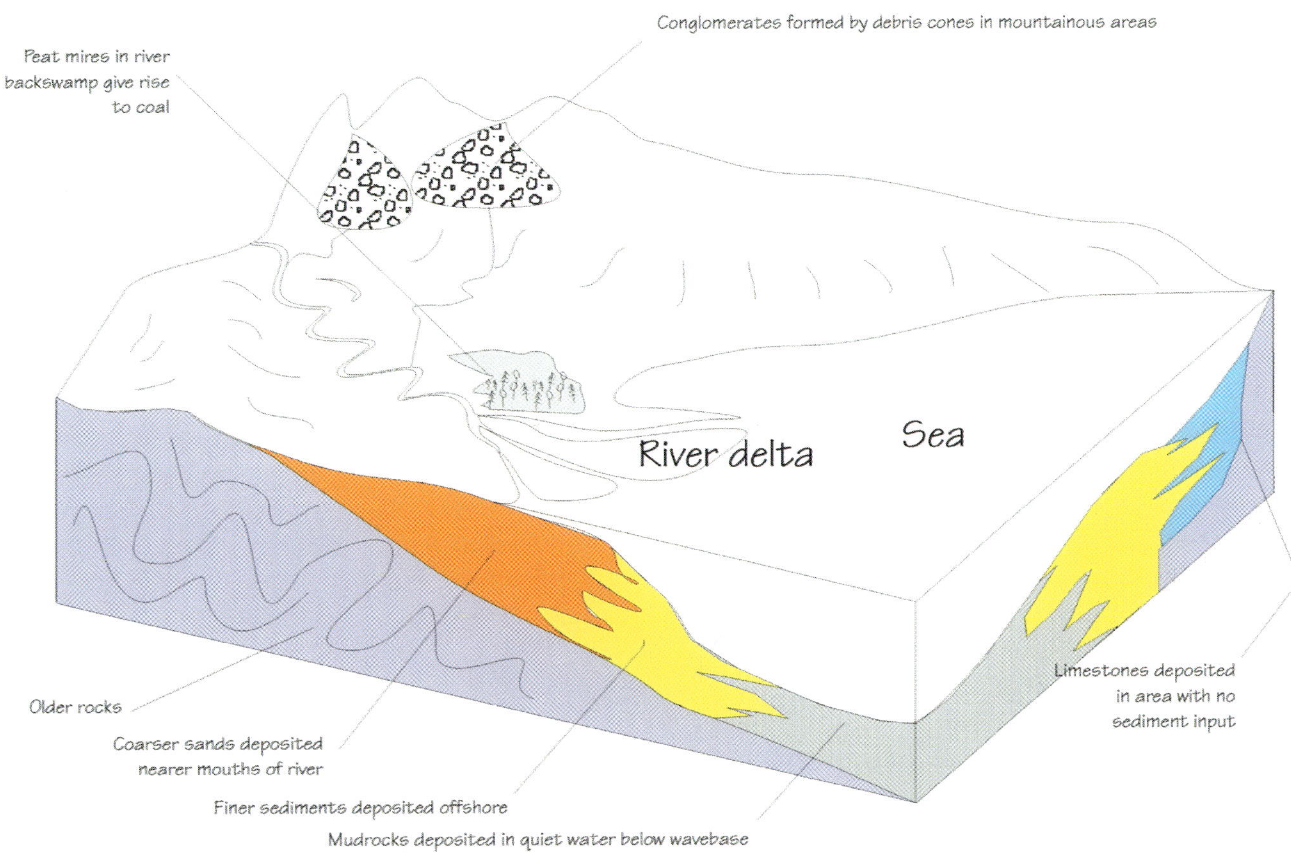

Diagenesis

Diagenesis is a process that acts on newly-formed sediments to convert them to rock. It normally includes compaction and hardening (induration), but may also include various chemical changes.

Water-containing sediments are compacted and much of the water is squeezed out. Acidity levels may change so that oxidation or the opposite, reduction, may occur. Limestone may be dolomitised (some of the calcium is replaced by magnesium to produce the mineral dolomite instead of calcite), and loose sand grains may be cemented. Nodules may form as elements are preferentially concentrated within the sediment, and buried remains of organisms may be leached away or be replaced by hard minerals. Buried peat is greatly compressed and begins its transformation to coal. Large organic molecules are degraded and begin the transformation into possible petroleum. Sediment grains, such as felspar and clay minerals, are altered to more stable minerals.

There is no firm boundary line between diagenesis and metamorphism (q.v.).

Sedimentary structures

Sedimentation is not usually constant in rate or source; sediment does not neatly settle to the bottom like mud in a jam jar of water. Depositional environments are usually dynamic. The deposition of any sediment is governed by the nature and flow of the transporting fluid and its later movement within the sediment; gravity; and biological and chemical changes. These can lead to the formation of sedimentary structures. Close analysis of these forms reveals details of the depositional environment. The following can only be an abbreviated account.

Sedimentary structures range in size from centimetres to kilometres in scale and are the record of the processes that deposited the sediment. They can be used to deduce palaeogeography and hence to predict the location of oil reservoirs or other features, as well as to determine whether strata have been inverted.

Sedimentation produces layers, or beds, of varying thickness. Beds less than 1 cm thick are termed laminae. Many sedimentary processes vary cyclically and can produce beds or laminae of contrasting type, perhaps on a daily basis or tidal basis (as flow velocity falls, finer, suspended particles may fall to the bottom). Currents may oscillate with the tide, or be regular in direction, producing ripples and related structures of different size and symmetry. These 'bedforms' (Figs.37 and 38) range from ripples about a centimetre high to 'dunes' several metres high and many metres in wavelength. When these features are formed in environments such as

Fig.37 (top) Sole markings (flute casts) on inverted slab of Carboniferous sandstone, north Cornwall; inverted ripple marks on slab beyond. Fig.38 (above) Close-up of ripple marks from Fig.37; yellow film carton gives scale. This is the underside of the sandstone bed, which has therefore been inverted by earth movements.

river channels, they may interfere with earlier bedforms, giving rise to cross-stratification (Figs.39, 40 and 41). Whole river channels may change course and cut into earlier channel deposits or peat beds, resulting millions of years later in impersistent coal seams. Interruption of coal seams in this way is termed a 'washout', and can be so extensive as to be economically very significant. Such channels can also be eroded below sea level into existing but unlithified deposits (Fig.42).

Fig.39 (top right) Cross-stratification, Ordovician sandstone, Shropshire: note how bedding is cut into by later, curved, bedding above hammer. Fig.40 (below) Oölites, because they are formed of rounded limy grains, can be cross-stratified too: examine the thick central set of beds in this view, above the obvious horizontal line. The cross-stratification (parallel sloping lines) dips to the right. Leckhampton Hill, near Cheltenham. Fig.41 (bottom left) Weathering on this mass of highly-exposed sandstone has picked out the fine detail of the bedding. Upper Coal Measures, South Wales. Fig.42 (bottom right) At the top of this quarry face in hard Ludlovian siltstones, a distinct curve can be seen where a submarine channel has cut into slightly earlier deposits. Shropshire.

In some circumstances, such as undersea slumping (like a landslide, but underwater) sediments may form a 'density current' that suddenly introduces a great volume of sediment to an area that otherwise receives restricted sediment input. The mechanisms of flow produce characteristic sedimentary structures. A turbidity current, in which sediment is kept up by the upward component of the flow turbulence, tends to deposit graded bedding, normally fining upwards from coarse sand to mud-grade particles.

Fig.43 Upper Carboniferous turbidites, north Cornwall.

Sediments of this type are often deposited in deep ocean trenches and basins with relatively unstable surroundings, and are found in many of Britain's areas of older rock. They tend to be repetitive, giving rise to alternating sequences of finer-grained darker shales or mudstones and paler sandstones (Fig.43), and are termed turbidites.

Fluid moving over a cohesive but unlithified bed may erode characteristic markings (Figs.37 and 38), which when lithified and later exposed reveal the direction of the current at the time of formation. This may in turn suggest the palaeogeography.

These markings are called sole marks. The presence of such markings is a way-up criterion, i.e. they are found on the base of beds and can be used to show (with other way-up criteria) whether strata are inverted. On sloping surfaces, sediment may suffer internal disturbance by localised slumping while still soft. The original stratification is retained but distorted, and may form a way-up criterion. There are numerous such; ripples and cross-stratification are also way-up criteria.

Another fairly common feature is desiccation cracking. This usually indicates exposure to the air, as of a mudflat, etc, (compare the cracking that the surveyor looks for in a garden when clay soil is suspected).

There are many other structures, of which space does not permit discussion, so that you should not be surprised to find unidentifiable features.

Ancient soils (palaeosols) are often found in certain sedimentary successions, notably in the Old Red Sandstone facies (Fig.44) of the Devonian and Upper Silurian, and in the Coal Measures, where root traces are often visible. In the Devonian, these palaeosols often form resistant and easily mappable features due to a concentration of resistant calcareous features formed around plant roots.

Bioturbation is a general term for the churning action of organisms on sediment. Worms, crustaceans, sea urchins, clams and other creatures either shuffle the surface around as they search for food, or excavate burrows for feeding or resting. Intensely bioturbated sediment may have completely lost its original bedding and appear homogeneous, or may show a confused pattern of burrows. Sometimes the burrows are clearly exposed in cross-section in a rock face. If you find a U-shaped feature, that is almost certainly a burrow. Recognisable burrows or feeding traces are trace fossils (see Chapter 7, Fossils) and can also be way-up criteria.

The concept of facies is very useful. It denotes the totality of features that together describe the overall characteristics of a sedimentary deposit; for instance, 'Old Red Sandstone facies' or 'graptolitic shale facies' or 'shallow marine facies'.

Fig.44 Palaeosols ('cornstone') in steeply dipping Old Red Sandstone, Devonian, west Pembrokeshire. Two palaeosols are marked by white bars, each about 700-1000 mm thick. The (originally) vertical lineations of calcified root pathways are obvious (parallel to the white bars).

Fig.45 (left) Hard, obviously fossiliferous (pale features) Carboniferous limestone, with pressure solution features, stylolites (the persistent subhorizontal marks); view about 1 metre across. Gower Peninsula.

Fig.46 (below) Penrith Sandstone (with reddish cement tinted by iron compounds): the grains are rounded and this, together with other factors, suggests formation as desert dunes.

Pressure solution may be found in carbonate rocks, where irregular lines, which are usually sub-horizontal with reference to the original bedding, cross a rock face. These are stylolites. They are usually clearly marked because they contain insoluble residual clay (see photograph of hard limestone, Fig.45).

Siliciclastic (or terrigenous detrital) rocks

Siliciclastic rocks are formed of accumulations of eroded particles of silicate minerals in the main, hence the name. A complex system of classification exists, referring to grain size, matrix type (the material between the larger grains), grain type, etc. However, for simplicity they are reasonably easy to consider simply in terms of grain size, with a continuous variation ranging from conglomerate (large pebbles) through coarse, medium and fine sandstones, then through silt to mudstone/clay.

In general, the larger grain sizes are found close to the point of origin; as the grains are eroded and transported they become finer. Consequently, the presence of very coarse sediment at a particular location suggests that high ground, a cliff line, or a fault scarp may have existed nearby at the time the rocks were laid down. Conversely, the presence only of uninterrupted clays or mudstones suggests that the shoreline lay far away.

Siliciclastics may be variably cemented, usually by silica - a hard cement - but also by calcite (like limestone), iron oxides or even weaker materials. The study of these elements of the rock suggests how it was formed. For instance, the presence of very rounded grains in a sandstone is normally taken to indicate a former desert environment, as in the case of the Penrith Sandstone (illustrated right, Fig.46).

Fig.47 Milford Haven, Pembrokeshire: a whole geology lesson in one place. The conglomerate, on the left, shows whitish quartz pebbles and whole rock fragments of several types. To the right it is overlain by medium– to coarse-grained sandstones. The whole arrangement probably developed as a fan adjacent to a steep slope formed by faulting. The rocks have subsequently been tilted about 70 degrees out of the horizontal by folding. The picture is complicated, towards the top, by faintly pinkish quartz veining. Coin gives size.

Conglomerate and breccia

At its source, sediment may consist of large fragments, of pebble size or greater. These may be preserved without significant further attrition to form a lithified sediment. If the individual clasts are rounded, it is conglomerate (Fig.47); if rough and angular, breccia. The presence of such rocks suggests the former existence of a major relief feature, such as a cliff or fault scarp, nearby and the dimensions of the conglomerate body, if they can be located reasonably well, can locate such a relief feature quite closely. Conglomerates and breccias are usually resistant to weathering (but see below) and often form useful marker horizons for mapping purposes, although their overall extent is usually limited. Some conglomerates may contain recognisable clasts of older rock, allowing the stratigrapher to draw conclusions about past events. If a rock contains pieces of another recognisable rock, then it must be younger, and so on.

Sandstone (Figs.41 and 46)

Sediment with particles in the size range 62.5 to 1000 microns (μm) is classed as sand. If lithified, rock with such particle size becomes sandstone. The sandstone may be classified in various ways, including by type and proportion of sediment grains, proportion and type of cement, etc. Hence a ferruginous sandstone would be significantly iron-stained and brownish.

An important type of sandstone is greywacke (Fig.48 opposite), which contains a high proportion of less stable minerals. Such sediment is thought to have been eroded and deposited fairly rapidly, before the non-quartz grains were destroyed by attrition or chemical processes. Such rocks are found predominantly in upland parts of Britain, in Cambrian to Silurian rocks formed during periods of active mountain-building and rapid erosion.

Most sandstones consist almost entirely of quartz, in which case their resistance then depends on the qualities of the matrix.

Mudrocks: siltstone, clay, mudstones and shales (Figs.34, 49 and 50)

Approximately 50% of all sedimentary rocks fall within this group.

Sediment particles in the range 4 to 62.5 microns are termed silt. The division is arbitrary. Lithified silt is a siltstone. Particles below 4 microns are clay. Mudrocks are usually poorly exposed because they weather readily to soil which becomes vegetated. As they are fine-grained, they are difficult to study. Mudrocks may be apparently massive (mudstone), or may be clearly laminated and/or fissile (shale). Fissility usually results from preferred orientation of fine particles, such as mica and clay minerals, parallel to the bedding.

Mudrocks are usually soft and relatively or highly impermeable, although they can often take up (but not transmit, hence the impermeability) massive quantities of water by adsorption onto the atomic lattices of the constituent minerals. Poorly indurated mudrocks may be little more than clay (fig.49), while mudrocks that are old and/or have been deeply buried in their history may be remarkably durable. Hard mudrocks may be termed argillite.

Mudrocks may contain nodules, typically under a metre in size, of ironstone, pyrite, calcareous concentrations, etc, which have formed by preferential aggregation of minerals. The latter grade into nodular limestones. Sedimentary ironstones are often common in some formations.

Because of their weakness and impermeablity, mudrocks shear more readily under stress and are often implicated in landslides.

Fig.48 Interbedded turbidite greywackes and shales, Silurian, north Pembrokeshire; note poorly-developed slaty cleavage in the shales, above the baseball cap.

Fig.49 Soft, unstable London Clay cliffs in the Isle of Sheppey.

Properties

To the surveyor, the mechanical properties of the above rock types may be the most important aspect. For instance, a conglomerate full of hard quartz pebbles may be, overall, quite weak compared with a nominally weak mudstone that has been deeply buried and hardened and is very cohesive. Even so, as a gross generalisation, conglomerates and sandstones normally form resistant landscape features and can stand as cliffs. These rocks can usually be identified fairly readily in the hand specimen. However, it is important to bear in mind that a rock that forms a strong cliff at one point may be weak and readily erodible only a few hundred metres away, due to differences in the cement quality, and so assumptions as to resistance and foundation quality should always be made with caution. This is particularly true, for example, of the sandy parts of the 'Lower Greensand' of the Sussex Weald.

Carbonates ('limestones') (Figs.45, 50, 51, 52 and 53)

Carbonates have in common only that they consist predominantly of calcium carbonate (mostly as the mineral calcite) and/or (less commonly) magnesium carbonate. By comparison with clastic rocks, carbonates are more

Fig.50 Persistent tabular argillaceous limestone interbedded with muds, shales and marls; west Dorset.

reactive chemically and undergo more changes during diagenesis. Few limestones are entirely free of magnesium carbonate, which normally appears as the mineral dolomite. However, some limestones may be extensively dolomitised and may be noted as such on maps. In limestone-rich areas, quarries or in some cases mines may be expected. In limestone-poor areas, any local limestone development may have been intensively worked in the past for agricultural or building lime and in such case its outcrop may be marked by a series of old quarries, depressions, pools, etc.

Limestones may form in a variety of environments but are predominately marine in origin. They are often of organic origin and may be highly fossiliferous (Fig.51) as well as consisting of a mass of compacted organically-processed pellets. Like the siliciclastic rocks, they have

Fig.51 (right) Shelly limestone, middle Jurassic: eroded shell fragments winnowed by tide and waves. Fig.52 (below) Oölitic limestone: the rounded grains or oöliths are clearly visible.

highly variable qualities. In terms of hardness alone, for instance, some carbonates, like the hard, grey limestone shown, are almost impossible to break with a hammer except at thin edges, and can be quarried in large masses; while others are quite soft, or nodular or thinly-bedded, like the argillaceous limestones of the British lower Jurassic (Fig.50) that have been much used for cement manufacture.

Oölitic limestones are common in parts of the English 'stone belt' that stretches from Dorset to Lincolnshire and includes the Cotswolds. Tiny grains, rather like mustard grains (Fig.52 previous page), are visible to the eye. Such limestones have formed valuable building stone for centuries and feature in many of the more famous British buildings. Bath stone and Portland stone are famous examples, but there are many others.

A famous controversy arose when the modifications to the British Museum were carried out. Portland Stone was specified, but a French limestone, *Anstrude Roche Clair*, from Yonne, was actually used. This is similar but by no means identical, as it is more beige in colour than the near-white Portland. In terms that the reader will understand, the specification referred to 'an oölitic limestone (Portland Limestone from the Base Bed or similar) to BS5390'.

Chalk (Fig.53) is a well-known British limestone of Upper Cretaceous age, which is widely distributed over southern and eastern England, with minor outcrops in Ulster. It appears more or less uniform to the casual eye, except for the prominent layers of flint in the upper parts, which are the source of so much gravel and flint walling. In detail it displays numerous minor variations including muddy (marl) layers and bored erosion surfaces. It is composed largely of minute calcareous plates (coccoliths) from microscopic plants, the *Coccolithophorida*. These flourished in a warm, sediment-free sea and sank to the bottom on death to give a rock that consists almost completely of calcium carbonate.

Fig.53 Chalk: not as uniform as is often thought.

Chemical precipitations

Evaporites are formed by the evaporation of more-or-less enclosed bodies of salts-laden water (usually seawater). They are not exposed at the surface in Britain because they are highly soluble. As erosion lowers the surface of the land, so the evaporite, as it were, retreats underground. Complete evaporation of about 1,000 metres' depth of sea water (at today's salinity levels) would produce about 15 metres of evaporites. Evaporites contain a sequence of different minerals: calcite and dolomite, gypsum and anhydrite, halite (rock salt) and various potassium and magnesium sulphates and chlorides. In practice the full sequence is rarely found.

In Britain, significant evaporite deposits are mostly confined to the deposits of the dry Permian and Triassic periods. They are found below the surface in Somerset, Worcestershire, Cheshire, and Yorkshire, with a history of significant extraction in the latter two areas. Late Jurassic/early Cretaceous evaporites provide the source for British Gypsum in Sussex.

As evaporites are easily dissolved, they can give rise to surface collapse even where they have not been worked. This particularly affects the Permian outcrop that runs from near Nottingham to Teesside. Solution cavities may also be filled by other minerals.

The section on Salt mining provides more information.

Ironstones

Ironstone is a sedimentary rock that contains more than 50% iron minerals (and is over 15% iron by weight). In the British Isles, most ironstones were formed under marine conditions, probably with low sedimentation rates. The minerals are usually hydrous iron compounds, and the rocks weather yellow to reddish-brown. Some are substantial enough to be used as building stone, such as the Jurassic Frodingham Ironstone, or the Claxby and Carstone formations of the Lower Cretaceous. Most ironstones are thin oölitic limestones with limited areal extent. In south Wales, haematite is found impregnating skeletal limestones within the Carboniferous Limestone.

Some ironstones are mudrocks impregnated with the iron carbonate siderite. They are common in the Jurassic, and if they form nodules or more continuous beds they are termed clay ironstones. Blackband ironstone is a dark, carbonaceous version. Such deposits have been extensively mined from the Carboniferous (where the extraction is masked by coal mining) and from the lower Cretaceous of the Weald.

Phosphates

Phosphatic sedimentary deposits are widespread but generally inconspicuous. Apatite, calcium fluorophosphate, is the commonest mineral. There is a strong correlation with erosive reworking of vertebrate (i.e. mainly fish) bone beds, with noted British examples at the top of the Triassic (the 'Rhaetic Bone Bed' and the upper Silurian 'Ludlow Bone Bed' of the Welsh Borderland. Occasionally deposits are of economic value, as with the Cretaceous Cambridge Greensand and Glauconitic Marl.

Coal, oil shale and petroleum

These form an important element in the British Isles.

Most organic matter is decomposed shortly after death and 'disappears'. This usually requires the presence of oxygen. Where oxygen is excluded, decomposition is incomplete or absent and the resultant matter can develop relatively stable organic compounds. Suitable environments include stagnant lakes; marine basins with stagnant lower waters devoid of oxygen; and swamps and bogs (lumped together as *mires*).

Coal

Organic matter accumulates in mires as a result of incomplete decomposition. This can occur at all latitudes. The resultant peat may be preserved as coal if suitable conditions occur. Ideally, some tens of metres of peat may be covered by later sediment as a result of marine transgression or delta channel switching, and will become gradually compacted to form coal. In Britain, most coal is thought to be derived from tree and leaf debris that accumulated in vast swamps during the Upper Carboniferous (or, in northern England and Scotland, lower to middle Carboniferous). These swamps formed on a low-lying plain formed by the large deltas of many rivers. Some coal (sapropelic coal) is derived from accumulation of organic material that is not in its position of growth, i.e. it has been washed into position, or is derived from floating algae, etc. Occasionally such material can form workable deposits.

Coal seams in British coalfields typically occur above sequences of strata that coarsen upwards in grain size, but this is not consistent.

After centuries of coal mining, extensive research has enabled British geologists to form detailed histories of the formation of the British coalfields.

Coals form a natural series, the 'coal series', which changes in 'rank', the term that denotes the level of coalification and increase in carbon content. This series runs: **peat** (with lowest carbon content) → **lignite** → **sub-bituminous coal** → **bituminous coal** → **semi-anthracite** → **anthracite** (with highest carbon) (Fig.54). Ultimately it terminates in **graphite** (pure carbon), which is not regarded as coal or coal-like but may be regarded as metamorphosed coal. Coalification is a mixture of diagenesis and metamorphism (see Metamorphic Rocks). For instance, in the western part of the South Wales Coalfield anthracite coal is associated both with greater depth of burial under later sediments and with major folds and thrust faults.

Fig.54 Anthracite: does not soil the hands.

Coals of lower rank contain more combustible gases, while anthracite is about 90% carbon. Coals of lower rank soil the fingers, while higher rank coals contain fewer friable components and may not stain at all. Higher rank coals tend to be blacker and brighter, with reflective fracture faces. As coalification progresses, methane may be released and this can accumulate within fractures and in adjacent strata. Methane is the dreaded 'firedamp', the cause of so many mine explosions.

Graphite, of course, is a different matter as it has been thoroughly metamorphosed to a different structure. Most British coals, certainly the massive Carboniferous coal developments, fall into the sub-bituminous to bituminous group, with significant quantities of anthracite. Later coals tend to be more lignitic, e.g. in the Bovey Tracey Basin of Devon. However, a shallow pocket of Eocene coal worked near Rochester may have been bituminous.

Sapropelic coals include cannel and boghead coals, which are derived respectively from altered plant debris, spores and algae, and from algae alone. Cannel coal is said to be so-called since it burns with a flame like a candle; it is known chiefly from Lancashire, while boghead coal, also known as Torbanite, is known chiefly from the Midland valley of Scotland.

Inorganic minerals are often found in coal in small quantities, chiefly sedimentary grains; nodules of siderite, dolomite, calcite and other minerals may appear; and pyrite is quite important and is significant in pollution terms. Coal balls, a sort of nodule, are thought to be early diagenetic products and may contain well-preserved plant remains.

Coaly streaks, that is, coal seams that failed to develop into significant thicknesses, are common in deltaic-type successions. The earliest British coal streak known to the writer is upper Devonian, but doubtless some researcher will be able to correct me. Seatearths, which are fossil soils, often with rootlet traces or indeed fossil roots, often occur without an overlying coal. Such seatearths may be highly siliceous, and are termed ganisters, while more argillaceous seatearths may be workable for refractory use and are termed fireclay.

Coal is most easily worked in areas where dips are gentle and faults are few. Most British coalfields have areas of poor ground with complex structures, however, especially in South Wales and the Somerset coalfield, where strata have been more intensely folded and faulted at the Variscan Front.

Although much British coal is now worked out, as far as past and current methods are concerned, considerable reserves remain at depth, in difficult areas, etc, such as under the Firth of Forth.

Oil shale

Oil shale (which can include siltstone and some limestones) contains a high proportion of organic material of which a good proportion can be extracted commercially by heating. As much as 50% of the weight of the rock can be extracted. Typically, such sediments have millimetre-scale laminae, alternately organic and clastic/carbonate. This probably reflects seasonal 'blooms' of plankton. As with coal, normal degradation of the organic content must be inhibited by the environment. In the Midland Valley of Scotland, a number of oil shale deposits are thought to have been formed in freshwater lakes.

Oil (Petroleum)

Organic matter included in sediments can, under certain conditions, become converted to petroleum. Processes of diagenesis and metamorphism are involved. Crude oil and gas are the two fractions, or types, of petroleum. The organic material, when buried to temperatures of 50-80°C, causes reactions which in turn cause the source rock to become 'mature'. Older rocks can reach maturity at lesser depths. The lower Jurassic rocks exposed in west Dorset are immature, while the same formations, where buried much more deeply in east Dorset, are the source rock for the Wytch Farm Oilfield.

The mobile oil and gas can migrate into overlying sediments until they reach the surface or are trapped in reservoirs beneath 'cap rocks'. Gas accumulates above the denser oil. Drilling through the cap rock makes the oil and gas commercially available. Most British oil is offshore, but significant quantities are, or have been, extracted throughout the British Isles.

Volcaniclastic sediments

The boundary between volcaniclastic and other sediments is blurred. In general, volcaniclastic sediments consist of grains of volcanic rock, arising from volcanic action that occurs at about the same time as the sediment is forming. Naturally, the type of magma available is a major influence. As the volcanoes themselves are positive features, they are often eroded away and the geologist eventually finds only those parts of the sediment that are peripheral to the volcano's area of influence. The small size of the ash or other material (and hence the available surface area) allows rapid weathering, while diagenesis is also rapid and can soon destroy original constituents and textures. The original erupted pyroclasts vary in size from fine ash (under 0.06 mm) to bombs (ejected while still liquid) and blocks, both over 64 mm. Volcaniclastic sediments can be divided as follows:

- Bombs → agglomerate
- Blocks → volcanic breccia
- Lapilli (2-64 mm) → lapilli-stone
- Coarse ash (.06-2 mm) → tuff (volcanic sandstone)
- Fine ash → tuff (volcanic mudstone)

Other volcaniclastic sediments include autobreccia, which results from fracture of the crust of still-flowing lava; and pyroclastic fall deposits. These are usually very well sorted, that is the coarsest material falls to earth near the volcanic vent, while fine ash may travel great distances

Sudden violent eruptions (such as that of Krakatau) can produce what are effectively instantaneous deposits of tuff (ash-fall tuff) over a large area. These deposits can form useful 'marker beds' for correlation over huge areas, especially where other date evidence is poor.

Other clastic or carbonate sediments with significant quantities of ash are tuffaceous.

Pyroclastic-flow and pyroclastic-surge deposits are formed by density currents of hot gas containing fine particles. They are produced by *nuée ardente* eruptions and can travel tens of kilometres on level ground. As they settle, pyroclastic flows become welded together by their own heat to produce an ignimbrite (older term: welded tuff).

'Drift'

Drift, or Superficial Deposits, is a 'dustbin' term traditionally used to include relatively thin sedimentary formations that overlie the 'solid' geology or bedrock and are often undergoing current depositional or transformational processes.

Weathering, erosion, and the production of sediment

Land that is exposed to the atmosphere and is not actively undergoing deposition of sediment is subject to denudation. This includes the total of all processes which conspire to destroy the land and to reduce it to its constituent parts. These mechanisms include weathering (including chemical and biotic weathering) and erosion by active agents.

There is no firm boundary line between weathering and erosion in terms of effect. Some erosion is catastrophic, for instance gullies that may be formed after flash floods, or sudden cliff collapse. However, much is imperceptible on the human timescale. An example is the weathering and gradual disappearance of the lettering carved into an exposed tombstone; after a few hundred years, any exposed inscription is unlikely to be legible.

The surveyor will be most familiar with weathering in its effect on building stones, pointing, etc. In many respects, the surveyor is well placed, therefore, to understand weathering intuitively. For instance, concrete roof tiles fail by gradual loss of the binding of their sand constituents, and become gradually thinner, while clay tiles fail by delamination as a result of freeze/thaw processes or chemical action. In similar fashion, bedrock exposed at the surface is attacked. More resistant rocks, such as granite or quartzite, will show little signs of weathering, as in the case of durable tombstones; while weakly cemented sandstones are readily damaged by the action of frost, gradually separating into the constituent grains. Rock which is out of sight below the soil, however, is also being broken down as a result of chemical action (including the action of plants and their products). If a foundation trench is dug, at the base of the soil profile is the deteriorated layer of bedrock which is the result of such breakdown. This renews the soil from below, while the top layer of soil is washed off by rain or removed by landslipping or other processes.

In short, the ground is being constantly reduced to fragments of varying size which can be carried away by active agents of erosion such as streams or glaciers. Glaciers are a particular erosive agent which, although now absent from the British Isles, have left their mark.

The results of the erosion of the landscape are often left more or less temporarily (in geological terms) on the bedrock surface to form superficial deposits. While these are generally quite thin, they often have considerable horizontal extent. Most buildings have foundations which are sufficiently shallow that they do not bear upon the bedrock, but rest upon the superficial deposits. It is important not become confused, at this stage, about the significance of what the geologist terms 'rock'. This term applies not only to the 'solid' rock, or bedrock, but also to the superficial deposits, including any loose, soft and deformable material, including unconsolidated sands and clays.

Surveyors should be aware that engineers may refer to such loose, soft and deformable material as 'soil' or 'engineering soil' as it can be excavated without drilling, blasting etc. It is therefore important, in discussion with engineers, to make sure what is referred to.

Setting aside the soil itself, which is the unconsolidated topmost layer of the ground that is subject to action by plant life, superficial deposits in the British Isles can be divided into alluvium, colluvium, and glacial or periglacial deposits. However, it is important to remember that the boundary between different types of deposit may be gradational or even impossible to define at all, while the geological map must show a line.

Alluvium

Alluvium (Fig.55) is clastic sediment, i.e. debris, which is deposited by streams and rivers. Where streams and rivers are very actively eroding, they deposit no alluvium, although at times of low water there may be small deposits of sediment in the river bed itself. However, as the river develops a flood plain, alluvium is deposited as part of this process. Any given body of alluvium may remain in position on the flood plain for a period ranging from years to thousands of years. Eventually, it will be caught up in a further movement of the river, will be re-eroded, and moved further down the river valley. This may occur several times before a given grain of sediment reaches salt water.

Alluvium is typically variable in lithology and properties, and even within the confines of a single building site, may include sand or gravel bodies deposited on the inside of river meanders, fine silt resulting from floods, mud deposited in abandoned meanders, or even pockets of peat. Consequently, its performance as the footing for a building cannot be guaranteed without investigation. The depth to rockhead (the surface of the 'solid' rock below) may be a matter of only a metre or so. However, it may also be tens of metres. This is especially the case near coastlines, since the rise in sea level since the last glaciation has allowed river valleys approaching the coast to become filled with sediment. This has been proven by seismic and borehole investigations in many locations.

Where thicker deposits of alluvial gravel have been laid down, they have often been excavated to provide building materials. This has resulted in many larger river valleys being a patchwork of landfill, remaining working pits, and water-filled pits. The latter are often re-used for leisure purposes such as fishing or water sports. Such a pattern of use can be seen, for instance, in the Thames and Trent valleys.

As rivers cut down into the landscape, they may leave stranded sections of former flood plain at higher levels. These are called river terraces. In some cases, the

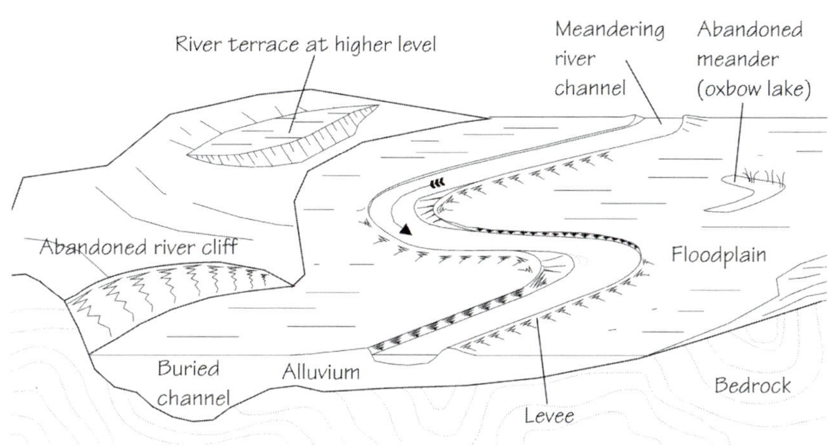

Fig.55 Anatomy of a floodplain.

sequence of the terraces has been thoroughly investigated, and the terraces have been numbered or named. This is of little relevance to the surveyor, and serves only to confuse the geological map, unless each terrace has a distinctive lithology and hence different performance.

Colluvium

Colluvium is the term applied to weathered or eroded rock debris that has been moved down the land surface by creep or by surface wash. It is generally quite thin, but in some circumstances forms a significant cover. In such cases it is often distinguished separately on geological maps as 'head'. Thicker deposits are often the result of frost action during ice ages. There is a special term for head formed on Chalk, namely 'coombe rock' which basically amounts to a mass of rubbly chalk. Colluvium may be indistinguishable in some cases from landslip deposits, and may grade imperceptibly into alluvium at the bottom of slopes. As with alluvium, it may be sufficiently thin that building footings can be cut through it to rockhead.

Landslides

Landslides and their results are dealt with separately in Chapter 9 and are not discussed further here.

Glacial and periglacial deposits

In round terms, it is only some 10,000 years since this country was subjected to direct glacial action. Glaciers are thought to have extended roughly as far south as South Wales and the Thames Valley, although the precise extent is still subject to much dispute (for instance, recent issues of a learned journal contained a quite heated correspondence and argument concerning the origin of a certain hollow on Exmoor). In fact, despite the current concern about global warming, it is very likely that we are merely between successive ice advances.

To the south of the area which was directly glaciated, and in the glaciated areas between glaciations, periglacial processes took place. These were dominated by frost heave, which resulted in cryoturbation, or frost disturbance, of softer materials and frost-shattering of harder materials. Many of the higher mountain peaks with rocky summits owe their jagged profiles to frost-shattering which still continues today. It is thought that the dry valleys of the British chalk lands were formed in periglacial periods when the ground was subject to permafrost, allowing normal river erosion to take place during periods of melting (alternative theories suggest that the erosion of these dry valleys could have been due to much wetter periods involving higher water tables). Such cryoturbation has been responsible for disturbance of many of the previously deposited sand and gravel deposits.

To the north of the purely periglacial area, the land was subject to several successive glaciations. At least two, probably three, and maybe four main glacial advances (and many lesser advances and retreats) took place over a period of several hundred thousand years. The growth of the ice resulted in major erosion of highland areas in particular, resulting in much in debris which was then transferred by the advancing ice to lower areas. As the ice sheets melted, the debris which they had eroded from the highlands was redistributed by river action. This has resulted in complex sheets of various thicknesses of sandy, gravelly and silty deposits.

Only recently have studies of glacial processes advanced to the point where some of these deposits can be satisfactorily divided into different ages and causes. There is a widespread deposit termed till (Fig.56, see opposite page), which was formerly usually termed boulder clay (a term to be found on many maps). Typically, this is neither sorted nor stratified, but consists of boulders and pebbles of varying sizes supported in a matrix dominated by clay-sized particles. The boulders usually reflect the geology of the country which has been crossed by the depositing glacier. Such deposits may be found at quite high altitudes, reflecting the ability of ice sheets to rise over hills and interfluves.

The till may be divided into various types (for instance, depending upon method of emplacement) which for the purposes of the surveyor are largely irrelevant. It is more important to note that the river banks, slopes, etc, formed in till or glacial deposits are commonly relatively unstable, and that small changes in their environment

Fig.56 Till and sands exposed in a sea-cliff in the Lleyn peninsula, North Wales.

(such as a temporary excavation) may increase this instability to a critical point.

Glaciofluvial deposits are also quite important. Glacial sediment is redistributed by water (including within or under the glacier) and can form linear features (eskers), shapeless mounds, obvious deltas (where a lake is present) and so on. Again, these features are usually unconsolidated.

Areas of glacial deposits often include hollows, formed by phenomena such as the melting of isolated lumps of ice ('kettle holes'), which subsequently become filled with unconsolidated deposits laid down in small lakes or ponds. These add to the high degree of lateral variability of the glacial deposits generally, and reinforce the need for adequate subsurface investigation when new buildings are to be constructed.

In some locations, the presence of the glacier ice, or of glacial deposits, has formed lakes which during the time of their existence gave rise to fine-grained deposits which now form dry land.

Older geological maps, in particular, may distinguish poorly between different types of glacial deposits. However, variability is generally such that any lithology indicated by the map should be taken only as a starting point for consideration.

Many surveyors operating in the southern part of the British Isles will never have cause to become concerned about glacial deposits particularly. However, those operating in more northern parts of Britain, including the heavily developed areas of Lancashire, Yorkshire, Tyneside, Ulster and the Midland Valley of Scotland, may find glacial deposits significant on an everyday basis.

Metamorphic rocks

As noted, there is no firm boundary between diagenesis and metamorphism.

Metamorphic rocks are formed by the transformation of pre-existing igneous, sedimentary or earlier metamorphic rocks under the action of heat, pressure or both (Table 4). Typical metamorphic environments would be deep under a developing mountain range, or next to a volcanic mass. Affected rocks recrystallise and new minerals are formed. The overall chemistry of the rock may be largely unchanged, or elements may be added or removed. The most intense processes produce rock that barely resembles its progenitor. Ultimately, some silicate rocks may be granitised and be virtually indistinguishable from primary granite.

Table 4

Typical metamorphic transformations

Parent Rock	Metamorphic Product
Granite	Gneiss
Andesite	Amphibolite
Mudrock	Phyllite, slate
Limestone	Marble
Sandstone	Quartzite

Conventionally, metamorphism has been classified as:

- Regional metamorphism.
- Contact (or thermal) metamorphism.

Regional metamorphism has affected large masses of rock, mainly within orogenic belts. The affected rocks are commonly deeply buried, and are not only heated but are subjected to substantial stresses. This leads to rocks which are often strongly foliated (i.e. they have a more or less laminated texture) such as slates, schists

Fig.57 (top left) Mica schist, with reflective surface formed by fracture along planes formed by mica crystals. Fig.58 (above) Typical Lewisian gneiss, with prominent bands of pink felspar. Fig.59 (top right) Garnet-biotite-gneiss: polished surface. Fig.60 (right) Basic gneiss, Midi-Pyrénées, France. Ringed object is a light meter for scale.

(Fig.57) and gneisses (Figs.58, 59 and 60).

It is quite commonplace for older rocks to be re-metamorphosed on more than one occasion, giving rise to a complex pattern of distorted foliations and a history of recrystallisation that is difficult to follow. As a result of this deformation, the minerals often recrystallise in alignments which reflect the directions of stress applied to the metamorphosing rock.

Commonly, the platy minerals, notably micas, are aligned at right angles to the maximum shortening direction and this produces a texture called schistosity (Fig.57). In intensely deformed rocks many minerals can form aligned flattened grains. Original banding or veining can be distorted, kinked and sheared in various ways (Fig.61, see opposite page).

Contact metamorphism is more localised. It is not accompanied by substantial stresses and the rocks which have been affected are commonly unfoliated. The heat from rising magma intrusions within the crust is transmitted to the adjacent rocks, and this gives rise to a series of concentric belts (in the ideal situation) of recognisable mineral suites which surround the intrusion with an obvious and regular pattern that can be delineated on a map.

Fig.61 Foliated granite, with quartz veins kinked by shearing. View about 1 metre across. Pyrénées Orientales, France.

The minerals in such rocks grow in random directions and this often gives rise to a very tough rock that does not split in any preferred manner, and this is termed a hornfels.

The metamorphism affects different rocks in different ways. For instance, mudrocks are often very reactive, because the original constituents, mainly clay minerals, have relatively limited stability. This permits them to develop a sequence of minerals which can be used, almost like a thermometer, to deduce the temperature and pressure at which metamorphism took place.

It is thought that metamorphism takes place in a near-solid condition. Any pore fluids present in the existing rocks are thought to play only a minor role, so that most metamorphic rocks retain many of the characteristics of the parent material. Although extreme metamorphism can lead to the onset of partial or complete melting and the formation of igneous rock, in most places metamorphism has not been so extreme and it is often possible to trace the bedding of original sediments.

Metamorphosed sediments are called metasediments, and the finer grained types are called pelites. Coarser grained metasediments are called psammites.

Metamorphic processes of more localised importance include:

- **Hydrothermal metamorphism**, whereby circulating hot water causes hydration of existing minerals and the formation of veins which are in equilibrium with the hot fluid.
- **Cataclastic metamorphism** is localized in fault zones and denotes fragmentation and intense pressure with little or no heating.
- **Impact metamorphism** occurs at meteor or asteroid craters and includes formation of minerals stable at extreme pressure.

Nomenclature of metamorphic rocks

Because of the complexity of metamorphism, it has not been found possible to have a rigidly defined nomenclature system. However, the simplest system is to add the prefix meta- to the original rock type, as in metagranite, and metagabbro. However, for many metasediments there are already historic specific names, e.g. a metalimestone is a marble (Fig.62).

These descriptions can be qualified by indicating particularly obvious members of the mineral assemblage of these rocks. In very high grade metamorphic situations, it may be impossible to deduce much about the nature of the original rock type, since both sediments and igneous rocks can turn into a gneiss. *Paragneiss* is a term used to denote a rock thought to be of sedimentary origin, and *orthogneiss* (e.g. Fig.58) suggests an igneous origin.

Except for limited outcrops of largely pre-Cambrian metamorphic rocks in parts of the

Fig.62 Carrara marble, with saccharoidal (sugary) texture.

Midlands, North Wales and Devon and Cornwall, most metamorphic rocks in the British Isles are found in the areas of north-western Ireland and in Scotland north and west of the Highland Boundary Fault. The exception is slate, which is found in a number of locations in North Wales, the Lake District, and parts of Cornwall.

Slate

Slate (Fig.63) is particularly important in the field of roofing, as the reader will know. It possesses so-called 'slaty cleavage' (Fig.48), which is formed as described above by the recrystallisation of the parent minerals at right angles to the maximum stress. Carefully applied pressure can then split a block of slate to layers only a few millimetres thick.

Slate may be disfigured by small mineral inclusions, notably pyrite, and it may sometimes be possible to see traces of the original bedding as colour bands.

Fig.63 Disused slate quarry, Snowdonia: the prominent near-vertical trend is the slaty cleavage. The trend sloping at about 45 degrees is the original bedding. Tiny figures to left give scale.

Chapter 5
Structure - the attitude of the rocks

Faults and folds

How do rocks lie in the ground?

Leonardo da Vinci was one of the earliest to deduce that rocks that had once been below sea level had been uplifted – they contained recognisable marine shell fossils.

The earth's surface is a restless place. On a day-to-day basis this is only signalled to us by the occasional catastrophic earthquake, volcanic eruption or landslide. On the scale of a human lifetime, movement is not apparent; but North America is drifting away from Europe at 1-2 cm per year, the North Sea is subsiding and parts of north-west Scotland are rising. As the 'plates' that form the earth's crust move against each other, the rocks are crumpled, twisted, distorted, and sometimes heaved up into mountain ranges like the Alps or Himalayas. Where plates move past each other, features like the San Andreas fault, the highly destructive fault that passes through San Francisco, develop.

Few rock bodies lie exactly as they were deposited, and recognition of the structure of the rocks allows correct reading of the geological map together with better assessment of landslip risks. The principal structures are folds and faults, which have different types.

Folds are, in general terms, developed much as a carpet buckles if it is pushed against an obstruction. Folds or parts of them are illustrated in most of the figures in this section, from which it can be seen that a wide variety of styles is apparent, including zigzags (Figs.64 and 65).

Faults are fractures that represent adjustment by breaking, rather than distortion. These features

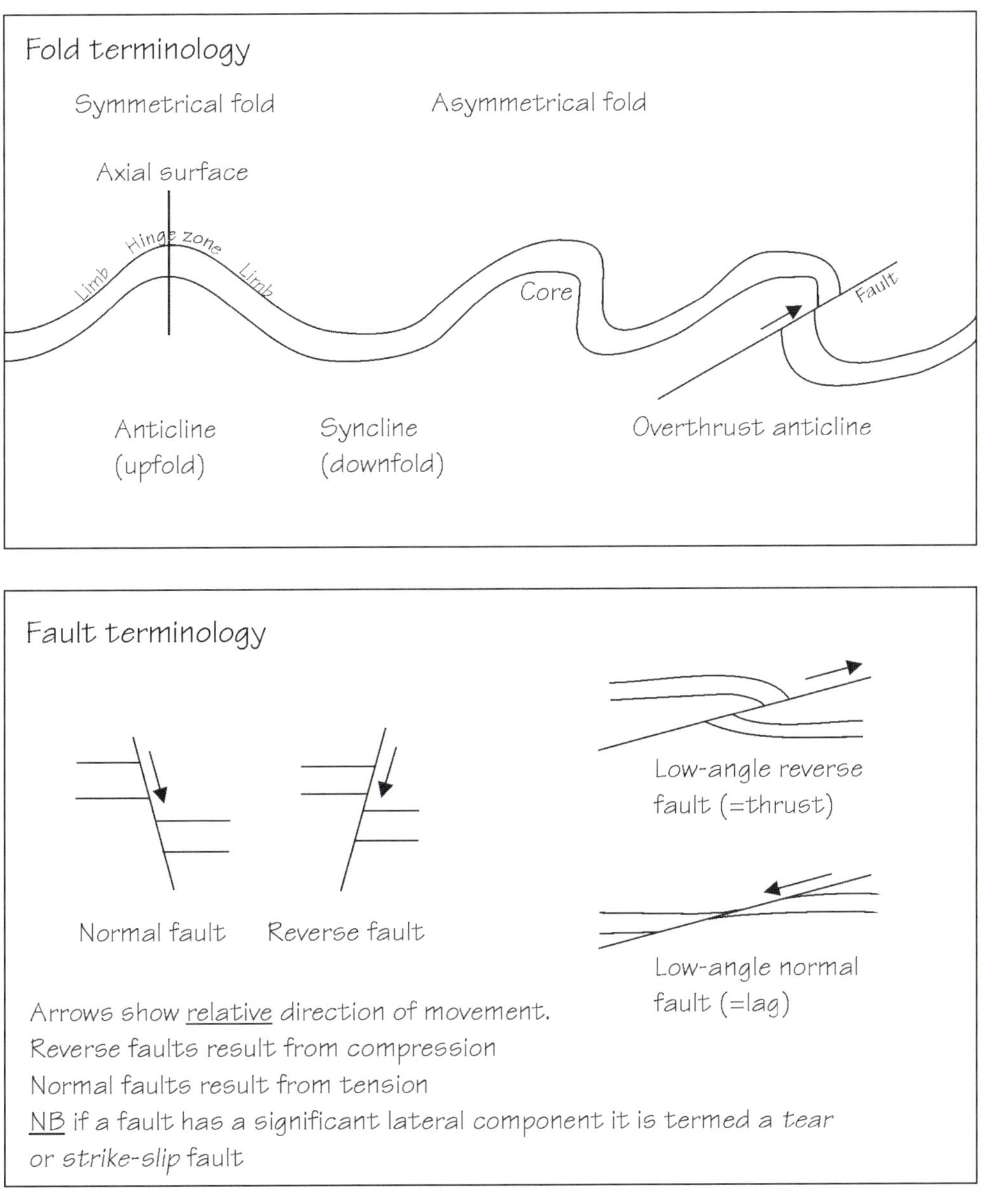

Fig.64 Folds and faults: basic terminology.

Fig.65 Complex zigzag folding in Eocene turbidites, Aragon, Spain. Fortunately most of the UK has much simpler structure than this.

typically are on the scale of 10 metres upwards. The displacement on a fault may be very small (microfaulting), or may be thousands of metres (vertical) or hundreds of kilometres (horizontal - strike-slip). For instance, the Great Glen Fault of Scotland separates land masses that, while now in contact, were once about 700 km apart according to some researchers.

On the smaller scale, rocks also feature joints and bedding. Joints are usually vertical or subvertical. They are developed on the scale of about 10cm -10 metres and represent local finer-scale adjustment to stresses of earth movement, shrinkage, etc, but without any appreciable loss of alignment, while bedding represents the original near-horizontal layering in sedimentary rock. Where bedding is widely spaced, rocks are said to be 'massive'.

Sedimentary rock that is not horizontal is said to dip, the dip (Fig.66) being the downslope direction. The direction at right angles to the dip is termed the strike.
Many textbooks tend to show idealised geological structures with sedimentary rocks neatly and evenly bedded. Occasionally, steeply dipping strata may still appear neatly bedded. In general, however, the more violent structures are often very untidy, with ragged and blurred fault fractures and perhaps a zone of crushed materials (Figs.67 and 68).

The attitude of the rocks is often reflected by their surface expression. For instance, the gently dipping limestones of the Cotswold Hills give rise to a series of flattish hilltops and a scarp some 200-300 metres high facing over the Severn Valley. Large areas of the Pennines feature alternating limestones and shales that form step-like topography. By contrast, the Hog's Back, in Surrey, is a long, narrow ridge formed on near-vertically dipping chalk but rises a bare 150 metres above sea level. However, structures may have little or indication at the surface (Fig.69).

Fig.66 (below left) These well-exposed bedding planes show their dip and strike clearly. Fig.67 (below right) Carboniferous limestone dips at about 80 degrees to the right. The beds in the centre have been violently kinked while the beds above and below are relatively undisturbed. Note the near-horizontal fault at the top of the tidemark; two other faults descend from the cliff top and run into the bottom of the kink. These rocks, at Tenby, are near the Variscan Front.

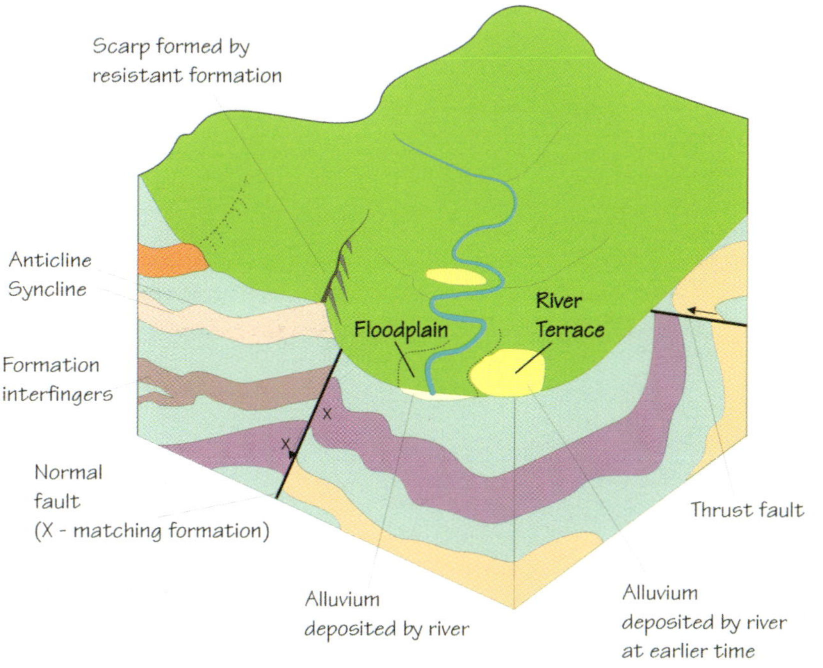

Fig.68 (left) This near-vertical bedding plane of Carboniferous limestone is broken by a complex fault with multiple fractures.
Fig.69 (right) Block diagram showing typical structures and relationship of some features to topography.

Faults can juxtapose rocks of greatly differing properties. It would be possible, for instance, for a thin sliver of clay to be squeezed between two sections of more rigid material such as limestone, giving rise to a small area of land with much poorer foundation properties. Furthermore, such a small mass might not have been detected and mapped.

Theory and reality

Textbook drawings show folds and faults generally as in the diagrams. Real rocks, however, can be squashed and twisted almost out of recognition, as shown in the illustrations, particularly Fig. 70. Here, the coal seam, which is relatively weak, has been compressed, squeezed and forced to adopt the volume that has been 'left over' following folding, faulting and crushing of the sandstones and mudstones above and below it. This sort of complexity can be barely hinted at on a 1:50,000 geological map. In areas that have undergone more than one major phase of earth movement, it may be almost impossible to decipher the sequence of events.

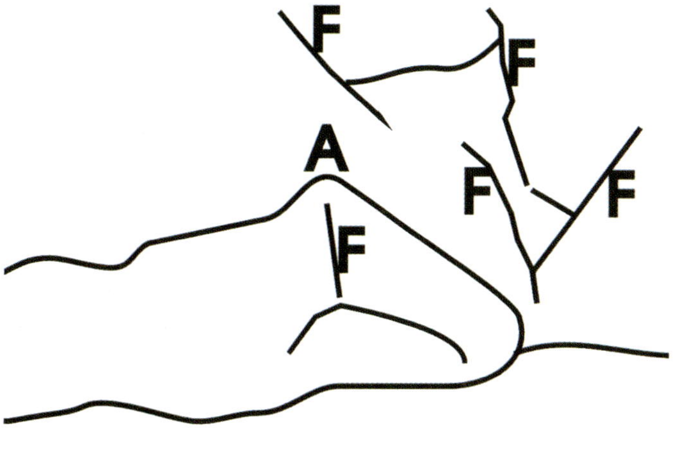

Fig.70 This contrast-enhanced view shows a section of Coal Measures comprising mostly mudstones and sandstones (stained by iron compounds). It features several obvious sections of fault (F), and an obvious anticline (A) with a sharp crest. Note how a coal seam (black) has been squeezed out of all recognisable shape and has been forced to flow into the space made available by the movement of the other rocks. This coalfield, in Pembrokeshire, was one of the first to be abandoned because of its complex structure. Height to point A from beach is about 3.5 metres.

Chapter 6 Earthquake hazard

Unlike California, Japan, or some of the Greek Islands, the UK is not thought of as seismically active. However it has a sufficient level of seismicity that more sensitive structures, such as dams and nuclear facilities, are potentially at risk. Major earthquakes are usually – but not always – associated with major crustal structures, especially subduction zones and other plate boundaries. This is why north-west Europe, in general, is seismically relatively quiet.

Earthquakes are recorded in terms of:

Magnitude: the amount of energy released, expressed logarithmically on the Richter Scale.
Most devastating earthquakes, such as the Kobe earthquake of 1995 (Magnitude about 7.2) and Izmit (1999) have had magnitudes of well over 6. Magnitude 5 would be high for the UK. Magnitude 6 and upwards generally leads to considerable destruction.

Intensity: reflecting the actual destructiveness and ranging from I – instrumental only, to XII – 'total destruction; objects thrown into air; ground rises and falls in waves'.

Recently, measurement systems such as the European Macroseismic Scale have been introduced but these have yet to be widely understood by the public. The destruction is influenced not only by the energy released, but by the rigidity of the rocks, and the depth of the earthquake, as well as other factors. The point on the surface above the earthquake focus is the epicentre.

In the UK, only a few earthquakes have entered public consciousness and these are largely historic. The 1884 Colchester earthquake was the most damaging in the last 400 years (estimated magnitude 5.2) and occurred in an area that has otherwise been almost devoid of earthquakes in recorded time. It damaged about 1,200 buildings but only a few people were killed. Certain areas have a pattern of repeated activity, such as the Caernarfon area in Wales (particularly strong shock of magnitude 5+ in 1984, with aftershocks), and upper Wensleydale in Yorkshire (significant quakes in 1768, 1780, 1871, 1933 and 1970). There is a constant recurrence in the Swansea area, with significant occurrences in 1727, 1775, 1832, 1868 and 1906. Such relative regularity suggests that quakes could recur in these locations.

Some seismically active areas can be correlated with major geological structures - for instance at Caernarfon and Swansea, major fault systems are present, which are thought to have been structural features for hundreds of millions of years. Some areas of activity are clearly related to the progressive collapse of old mine workings. Others, however, have no obvious cause.

The Dover Straits are the site of two of the largest British earthquakes (1382 and 1580 - estimated magnitudes about 5.75). Although only small shocks have been recorded since 1580, another significant earthquake cannot be ruled out. It is suggested, however, by some workers that the stress in this area has been released along structural lines in Belgium instead. Other recent earthquakes felt in the British Isles include Carlisle, 1979 (Magnitude 4.7), and Roermond (Netherlands) (5.9). The 1932 Dogger Bank quake is the largest recorded (6.1).

The areas of highest seismic risk in the UK, not surprisingly, are thought to be those where most activity has occurred in the past. It is estimated by the BGS that even in the higher-risk areas, if a building has a life of 50 years, there is only a 10% chance that it will receive shaking of intensity 6 ('trees sway and all suspended objects swing; damage by overturning and falling of loose objects') .

Even so, in areas of higher seismicity the risk should be pointed out, with the proviso that no certainty attaches to any figures.

The further risk arises, of course, that under a particular combination of circumstances, an earthquake could trigger a landslide.

Chapter 7 Fossils

Introduction

What is a fossil?

A fossil is the remains or trace (trace fossil, e.g. footprints) of an organism that has been buried in sediment naturally and subsequently preserved to some degree. Fossils are studied by palaeontologists.

Most organisms are preserved by the replacement of their body parts by minerals, often calcite (Figs.71 and 72) or pyrite (Fig.73, see p48), or by the formation of moulds or casts (Figs.74, see p48 and 75, see p49). Sometimes preservation can be amazingly delicate and detailed. For instance, a small dinosaur was reported from China in January 2003 in which feather impressions were found associated with all four limbs. Many organisms, however, are preserved only as thin films (such as the graptolites illustrated, Fig.76, see p49) and/or show such little contrast with the surrounding sediment matrix (Fig.77, see p49) that they are difficult to make out.

Fig,71 Small but beautifully formed: a Carboniferous brachiopod, Productus sp. Relatively easy to identify, and reasonably diagnostic (can be much larger than this specimen).

While the popular press tends to concentrate on dinosaurs, there are myriads of less spectacular, but equally interesting, types of fossil, some of which are often routinely well preserved.

NB: The illustrations here generally show items from the author's collection, not cleaned-up museum specimens, and represent what you may readily find in the field.

The use of fossils

Fossils are studied to shed light on evolution. They also allow relative dating of sedimentary and some metamorphic rocks (see below), and suggest an ancient environment for the deposition of those rocks, thus contributing to the entire geological jigsaw. New techniques such as DNA analysis (for some very recent fossils) and cladistics (a means of classification) have revolutionised some aspects of palaeontology. For most purposes, the surveyor will find fossils of little practical use. However, some easily recognisable fossils can be diagnostic of certain ages. For instance, in Great Britain a trilobite (see Fig.78, see p50) found *in situ* in the rock can only indicate a pre-Permian date. Consequently if a map showed rocks of later date to be present at that location it would be suspect. Similarly, if you find a belemnite (a Jurassic and Cretaceous fossil) in an area thought to be Carboniferous, again there may be an error in the mapping. Finally, the presence of *any* fossil is indicative of sedimentary (including some volcaniclastic) rock.

Although isotopic studies on graphitic rocks in Greenland suggest that the life forms

Fig.72 This bivalved fossil came out of the rock almost cleanly. Successive growth stages are apparent. It is a brachiopod belonging to the Rhynchonellida, a large taxonomic group that extends from Ordovician to Recent times. Many of them look almost exactly like this and can be used only by an expert for dating. Middle Jurassic; about 1 cm across.

on this planet were present 3.8 billion years ago, the issue remains open to dispute. Very few fossils are found below the Cambrian, and except for the immediate Pre-Cambrian they are generally extremely small. However, the surveyor is unlikely to be very much concerned with Pre-Cambrian rocks, which generally outcrop in relatively remote and rural areas of the British Isles.

To attempt even the broadest systematic look at the fossil world is impossible within the scope of this book. However, a few groups of organisms can be recognised by the complete amateur and yet have some use in dating, as shown above.

Most British fossils are invertebrate, that is, not having a backbone. All the animal fossils in the text figures are invertebrates. Fishes, reptiles, mammals and birds are not commonly found, or may be found as (say) shark teeth or fish scales only. Small fossils may not be immediately recognisable as such, and very small fossils (microfossils), such as pollen grains can be studied only in the laboratory by micropalaeontologists. Most macrofossils are in the centimetre-order size range (Figs.74 and 75). Do not expect to find *Tyrannosaurus rex* (to use a name now familiar to most) on your first outing! Until you have 'got your eye in' you may well miss well-preserved specimens which have to be carefully extracted from their rocky matrix (Figs.71, 74 and 75). On the other hand, you can turn over shale fragments all day and find nothing, even if you are a professor. Fossils are often concentrated in beds only a few centimetres thick. Until you identify these, you may find nothing at all, while the fossiliferous beds may be packed with fossils.

Fig.73 (below) An ammonite (rather like a squid, but with a coiled shell) preserved in pyrite (iron sulphide). Image about life size. Note rust filming on the specimen just visible at the top. Found only in Mesozoic rocks, but can be confused with other cephalopods. Fig.74 (right) Internal cast of lower Cretaceous snail: fossils of this general type are found throughout the stratigraphical column.

'Cheating' by visiting a site with an expert, or buying a suitable guide, is the easiest way to find and identify fossils quickly.

Many fossils, even if quite large, are almost impossible to recognise unless you already have some familiarity with them, although to an expert they may be perfectly obvious; while others may be easily dislodged whole from the rock (Fig.72) and be immediately recognised as biological. Many consist of broken fragments or partial remains only: for instance bivalved shells often become separated, while cast-off exoskeletal parts of trilobites and other arthropods (Fig.75) (which moult, or shed their skins, regularly) are usually found separately. Rocks may often be distinguished by the broad type of fossil that they contain, e.g. a limestone may be described on the map key as a shelly limestone, or a shale may be noted as graptolitic. In such cases, recognition of the precise organism is not necessary.

Trace fossils may be the footprints, resting marks, skin impressions, feeding traces or burrows of organisms. Often they cannot be related to a particular creature.

Fig.75 (top) An Ordovician trilobite pygidium (tail section): there is no sign of the rest, and this is probably a moulted section. About 0.5 cm wide. Trilobites in Britain are mostly Cambrian to Silurian in age; rare species persisted until Carboniferous times.

Fig.76 (centre) Graptolites: found only in Ordovician and Silurian rocks. Scale in mm.

Fig.77 (bottom) Plants from the Coal Measures preserved as indistinct coaly films. Scale in mm.

Fossil names (taxonomy)

Taxonomy is the study of the ordered arrangement of individual types of life forms. Organisms are identified uniquely by their genus (as in *Tyrannosaurus*) and species (as in *rex*). Note the use of capital and small initials respectively (often misunderstood), and italics. This is the Linnaean system of nomenclature. We are *Homo sapiens*, as most people know. There are higher groupings of organisms - families, classes, etc, - that are not immediately relevant. In any case, biologists, palaeontologists and taxonomists (specialists in classification) will often argue about the classification of organisms. Species can change their name or be transferred to another family, etc. Arguably, for instance, a bird is logically a dinosaur. There is also massive academic controversy about the relationships between the major groupings; none of this is important in the field.

Only experts can normally distinguish a species in the field (except for some particularly distinctive organisms), and often not then. If you can identify the genus, or even the family (the next higher grouping) in the field, you are doing well. Do not attach a specific name to any fossil: only a very few highly distinctive species can be relied upon for complete identification. Even experts may refer to fossils tentatively; for instance *Bloggsia* cf. *smithi* would be an organism similar to, but not certainly identifiable as, *Bloggsia smithi*. *Bloggsia*? or cf. *Bloggsia* would be even more uncertain.

Dating by fossils

Because organisms have evolved through time, there is a succession of varying types which are confined to greater or lesser timespans. Some organisms are too long-lived to be of use, while others, such as ammonites or graptolites, evolved rapidly, with species lasting for periods in the order of a million years. This has allowed palaeontologists to erect a system of zones to which rocks can be referred and thereby correlated with quite different rocks far away that nevertheless contain the same fossil types. The zone is the basic unit, but can be divided into smaller units and grouped into larger units, up to Period level.

In theory, zones are time-rock units, that is, the body of preserved rock that was deposited during a given timespan. In practice, organisms do not appear everywhere overnight, so that no zone or other fossil-based division can be truly of one timespan throughout its distribution. For most purposes this does not matter. Not only does the body of preserved rock have many gaps, small or large, but the time taken by an organism to spread is probably very much shorter than its lifespan as a recognisably distinct species.

Fig.78 Silurian trilobite, Diacalymene ouzregi, well-preserved in three dimensions; head to right. Moroccan; scale in millimetres. Commercial specimen about 60mm long..

Chapter 8 Stratigraphy of the British Isles

As with human history, the further one goes back in time the fewer are the preserved records. Additionally, many of the oldest rocks of the British Isles are in areas which are only sparsely populated. For these reasons, earlier periods of geological history are dismissed more cursorily than the more recent periods. Even so, in the following pages only the most generalised picture can be painted.

Reference to the geological timescale (Fig.4), the geological map (Fig.79, see overleaf), the terrane map (Fig.80, see p53) and the other figures is necessary for understanding. However, the geological sketch-map provided here is at too small a scale to give more than an approximate indication of the main distribution of both crystalline and sedimentary rocks. Likewise, the palaeogeographical maps are 'snapshots' showing conditions at dates 50 -100 million years apart, during which time whole continents can pass by. The glossary may also be useful. However, a few terms are best explained at the outset.

The joint lines along which continental masses collided are usually still recognisable features and are termed sutures (see Fig.80). Major faults that were formed during the same processes are also often still recognisable and both types of feature often separate areas with substantially different geological histories, termed terranes (note spelling!). When rocks are deposited on older rocks and there is a significant gap in deposition, this is an unconformity. The frequency of earth movement in the British area has given rise to numerous unconformities, which allow the position of earth movements in both time and space to be assessed. Thicknesses of strata vary greatly; the thickest accumulations are usually found in areas that subsided consistently over a long period, termed basins. When the sea rises relative to the land, that is a transgression; the reverse is a regression. The history of the rocks is worked out by the Law of Superposition. In short, if a given rock is cut by a fault, intruded by igneous rock, or overlain by another rock then that second feature must be younger.

It is important not to become confused between time-rock or chronostratigraphic units (see Table 1) and lithostratigraphic units. Lithostratigraphic units are mappable, lithology-based units. Chronostratigraphic units are bodies of rock that were laid down in a particular timespan.

The Old Red Sandstone (see p60) is a classic example: it was formerly regarded as Devonian in age, but it has since been realised that in South Wales it begins in the Upper Silurian, and its youngest beds are in many places dated to the basal Carboniferous.

Archaean rocks

Prior to the Cambrian, insufficient evidence exists to put together a fully coherent picture, although the oldest British rocks, in north-western Ireland and in north-western Scotland, show evidence of mountain-building from several billion years ago. It is clear that plate movements were taking place.

The oldest exposed British rocks are of late Archaean age and are found in the Hebridean Terrane in the north-west Scottish and Hebridean areas. They have long been known as the Lewisian Complex, although this is a shorthand term for a very complicated group of rocks which were much affected by deformation and metamorphism in a later pre-Cambrian time. Most of the rocks are gneisses which were once plutonic igneous rocks. There are limited areas of what are clearly metamorphosed sediments and erupted volcanic rocks. Uranium-lead dating of preserved zircons (zircon is a mineral, zirconium silicate, which is very resistant to alteration but is otherwise of minor importance) suggest an original formation date of about 3 billion years. High grade metamorphism probably occurred at about 2.75 billion years. The northern-most section of the Lewisian complex yields original formation dates of about 2.8 billion years, suggesting that two earlier terranes have been combined. Attempts have been made to relate the Lewisian rocks to possible subduction zones, etc, but this is tentative.

Inliers of the gneiss outcrop over widespread areas to the east of this Hebridean terrane.

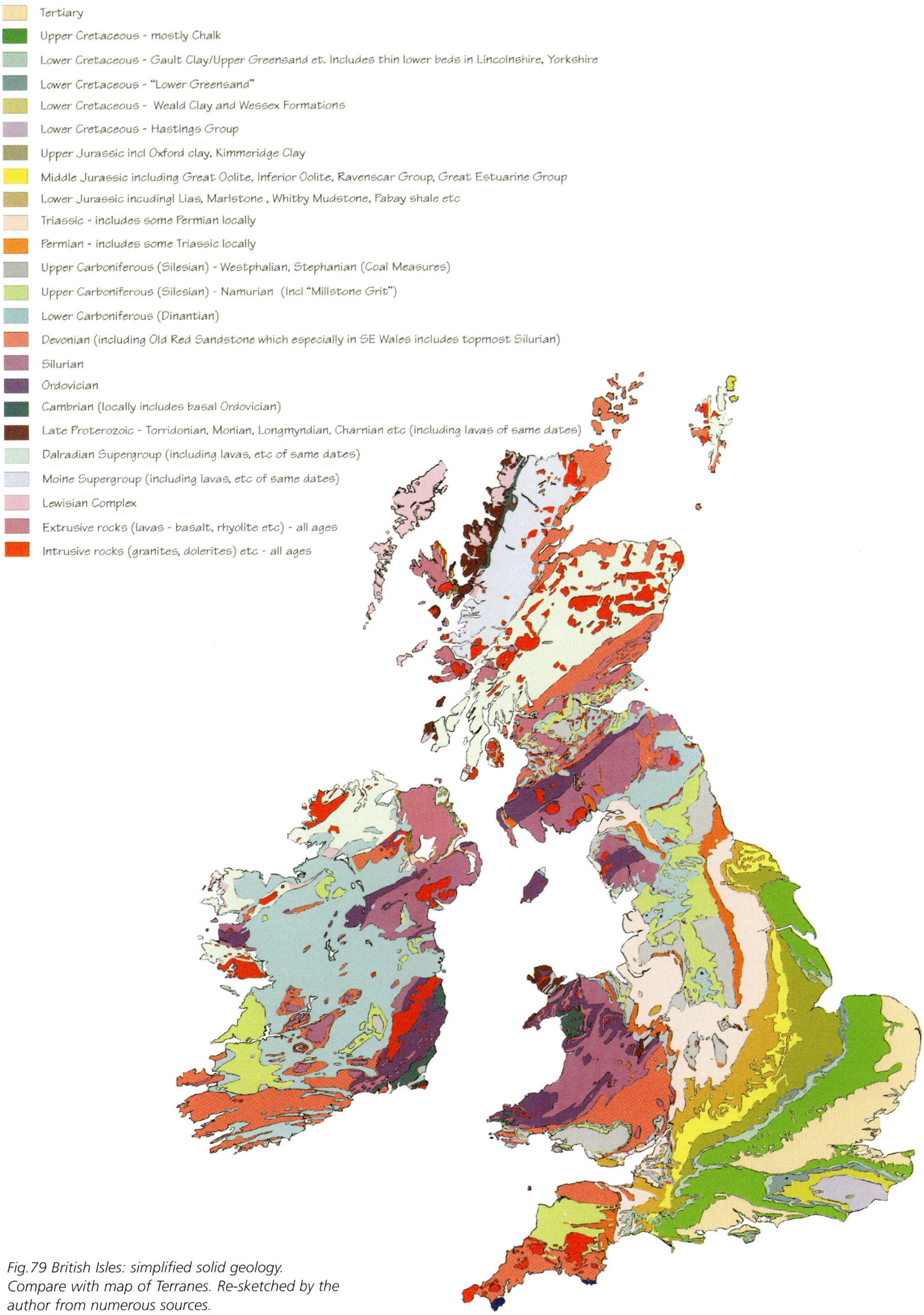

Fig.79 British Isles: simplified solid geology. Compare with map of Terranes. Re-sketched by the author from numerous sources.

The evidence suggests that by about 1.4 billion years ago, crustal extension gave rise to basins in which continental sedimentary rocks were deposited. These form the earliest section of the so-called 'Torridonian' succession, and include about 2 kilometres thickness of river, desert and lake deposits, mostly reddish. These rocks fill valleys in a hilly landscape of Lewisian rocks, and are thought to be about 1.2 billion years old.

The limits of the depositional basin are uncertain. By this time, continents were converging to form the supercontinent 'Rodinia', which included Laurentia, Siberia, Baltica and most of Gondwana. This gave rise to deformation and metamorphism in the basement rocks (basement is another shorthand term, for generally old,

Fig.80 The British Isles are made of a patchwork of previously separate landmasses: map showing major terranes and major fault and other structural lines. A number of small terranes are not shown, to avoid confusion. Redrawn by the author.

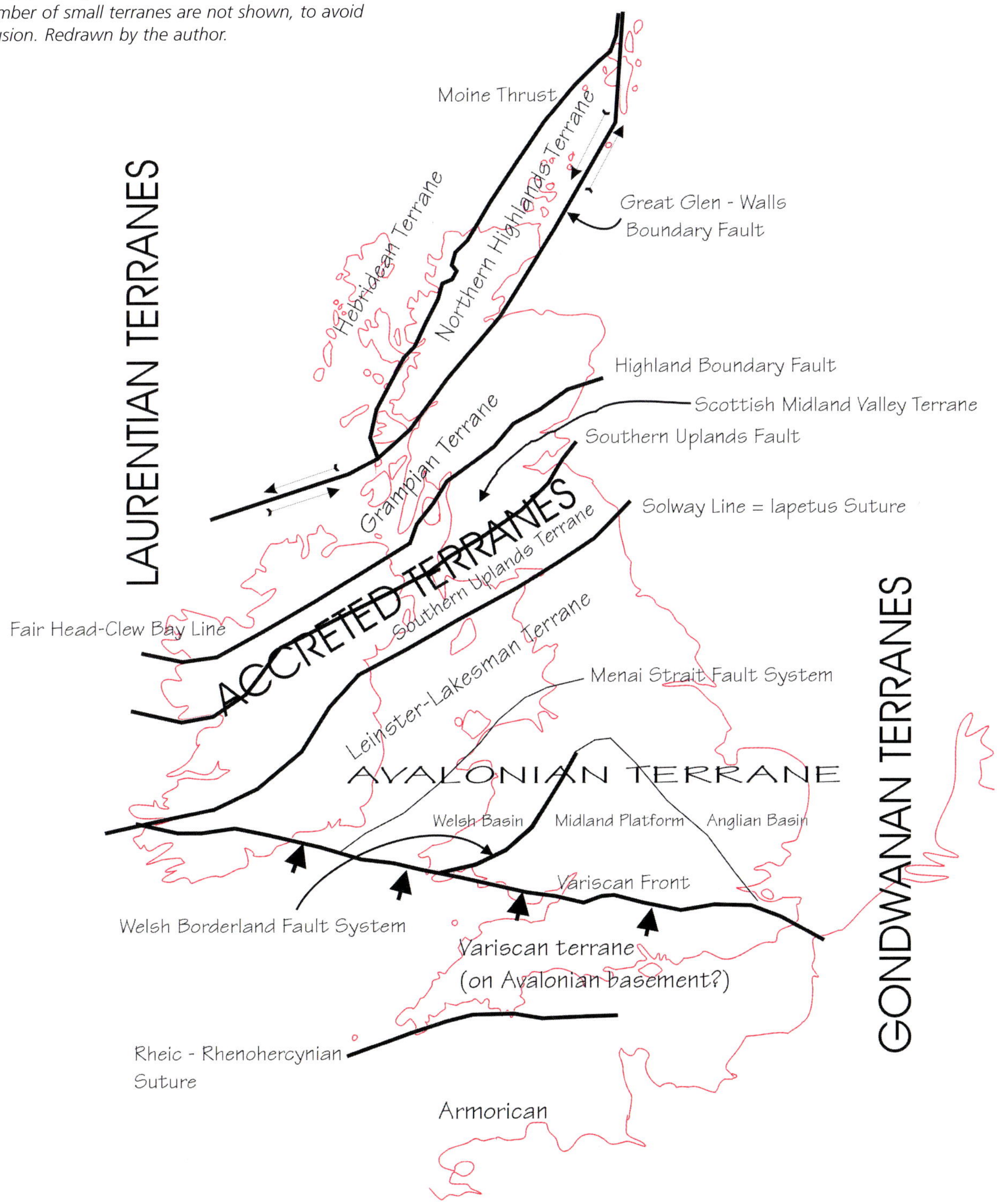

deformed and metamorphosed rocks on which a given younger succession of rocks lies).

The resultant mountains, the Grenville Mountain chains, gave rise to the accumulation of thick sedimentary deposits during the period about 1 billion to 800 million years ago, and these form the remainder of the Torridonian succession. This outcrops extensively along the north-west margin of Scotland and also in some of the Hebridean Islands. About 8 kilometres of mainly reddish sandstones cover an irregular topography of the Lewisian complex and the preceding Torridonian, with a relief of up to 600 metres. These sediments now form the strangely-shaped peaks of north-west Scotland such as Canisp and Suilven.

Again, the full extent of the basin cannot be assessed, but it is known to be enormous, at least 200 kilometres long.

All these rocks so far discussed form part of the former Laurentian continent.

Later Proterozoic to Silurian in Laurentia

To the present south-east of the Lewisian and Torridonian, much of northern Scotland is formed by the outcrop of the **Moine Supergroup**. This is a very thick succession of strongly deformed and metamorphosed sedimentary rocks between the Moine thrust zone (see below) and the Great Glen Fault forming the Northern Highlands Terrane. Sedimentation is approximately coeval with the Torridonian, but is not clear whether the Moine sediments were ever contiguous with the Torridonian and they may have been juxtaposed by later strike-slip movements of hundreds of kilometres. Moine rocks are also found in Shetland. A thickness of 5 kilometres or more is proposed for the supergroup. Despite the metamorphism, sedimentary structures are preserved and suggest a shallow marine environment.

To the south-east of the Great Glen Fault and the Central Highlands Terrane, there is a suite of very heavily metamorphosed sedimentary rocks, the Central Highland Migmatite Complex, which resemble parts of the Moine supergroup but show no obvious direct correlations. The largest outcrop of these rocks is to the south-east of Inverness. The junction with the Dalradian to the south-east (see below) is largely marked by a structural break, which adds to the mystery.

Both the Moine and the Central Highland Migmatite Complex show that further widespread deformation of metamorphism occurred during the Grampian Orogeny, during the middle Ordovician. However, there may have been significant late Cambrian metamorphism in addition.

The Moine and the Central Highland Migmatite Complex are intruded by a wide range of igneous rocks ranging from granite plutons to dolerite dykes, from a wide variety of ages.

In summary, this relatively small part of the earth's crust shows distinct signs of having been assembled by the coming together of various sections of supercontinents (the nearest modern equivalent is Eurasia), over a period from about 3 billion years ago to about 0.75 billion years ago.

By the later Neoproterozoic, about 750 million years ago, the geological record is slightly easier to interpret. Most of the Grampian Highlands (between the Highland Boundary Fault and the Great Glen Fault) and much of northern and western Ireland (north-west of the Fair Head-Clew Bay line) is occupied by the **Dalradian Supergroup.** Rocks of this supergroup, the total thicknesses of which add up to about 25 kilometres, range in age from about 750 million years to Middle Ordovician times. Consequently, Cambrian and Lower Ordovician fossils are reported, albeit sparsely.

The Dalradian comprises a sequence of marine rocks, mainly siliciclastic, but with a few limestones, and there are some important marker horizons. These include extensive basic volcanic rocks, and towards the top of the succession, fossil glacial deposits, the Port Askaig Tillite. The latter tillite has been implicated in the so-called 'snowball earth' hypothesis, the notion that there was a late Pre-Cambrian world-wide glaciation.

Much of the Dalradian has been affected by high grade metamorphism, but some parts, particularly in the south-west Scottish Highlands have not been so metamorphosed and are relatively easy to examine. In this area, and in north-west Ireland, it seems that the sediments were deposited on a basement not unlike the Lewisian complex. The rocks were probably deposited in a basin roughly parallel to the present orogenic trend, i.e. roughly south-west to north-east.

During part of the same period, marine sediments of Cambrian to Ordovician age were deposited unconformably on both the Lewisian complex and the Torridonian rocks, and are preserved as a long, narrow strip, parallel with the north-west Scottish seaboard. The present outcrop is thought to be roughly parallel to a shoreline to the north-west at the time of deposit. The succession suggests a marine transgression, with quartzitic sandstones in the lower section, and a general fining upwards to a more shaly sequence topped by the Durness Limestone Formation, the top of which is of Lower Ordovician age.

It is thought that the present area of the Moine Supergroup was probably overlain by a continuous blanket of Cambrian to Ordovician marine rocks, which would have been contiguous with the present Dalradian outcrop. However, all traces of this have been removed by later earth movement.

The Grampian Orogeny

A number of dynamic events occurred on the south-east margin of Laurentia with the Iapetus Ocean to the south-east.

The Grampian Orogeny occurred during Mid-Ordovician times and gave rise to a mountain chain that has ever since supplied clastic debris to the surrounding areas. It is though that the orogeny was caused by the collision of the Laurentian margin and a volcanic arc offshore to the south-east, which had developed above an intra-oceanic subduction zone within the Iapetus Ocean. The site of the volcanic arc is believed to be below the present Scottish Midland Valley. The original Laurentian crust is thought to correspond roughly with the area to the north-west of the Highland Boundary Fault and the Fair Head-Clew Bay line. The orogeny is thought to have lasted only for about 15 million years, since the oldest detritus from it is Mid-Ordovician.

To the south-east of the Highland boundary fault, there are no known or significant sections of basement rock. However, the geological story is continued by the **Middle Ordovician to Silurian** sedimentary rocks. These form much of the Southern Uplands and its continuation into the northern part of Ireland, and also outcrop in the Midland Valley of Scotland and its corresponding structural equivalent in Northern Ireland. The Grampian Orogeny had resulted in thrusting of part of the Iapetus oceanic floor on to the basement of the Midland Valley and this is now exposed as the Ballantrae igneous complex of South Ayrshire.

About 2.6 kilometres of sedimentary rocks, mainly siliciclastic, were deposited in deltaic and marine environments onto the complex. Included granite fragments in the sediments give ages of about 560 to 450 million years, suggesting erosion of an active volcanic arc to the north or north-west. This arc is now thought, as noted above, to be beneath the latest sediments of the Midland Valley.

A number of Silurian inliers in the Scottish Midland Valley and the Irish equivalent do not have their bases exposed. The general indicators are of regression through the Silurian with marine turbidite sequences passing up into shallow water sediments. Many inliers record an upward passage to fluvial sequences. In Ireland only, the Silurian rocks rest unconformably on older rocks brought to the surface by the Grampian Orogeny. Cross-stratification indicates derivation from an approximately southerly direction, suggesting the presence of rising ground to the south (see below). Correlation between the Scottish and Irish inliers is difficult.

The general synthesis is consistent with a progressive infilling of the Midland Valley basin.

Meanwhile, to the present south-east, in the Southern Uplands Terrane, a sequence of deep water Ordovician and Silurian sediments were deposited. The Ordovician begins with lavas and cherts, followed by black shales and turbidite sandstones. Volcanic clasts indicate that at one time sediment was being derived from an exposed landmass of Cambrian rock to the north.

The Southern Uplands Terrane is characterised by repeated thrusting and folding which verges (closes) to the south-east. The currently popular interpretation is that the pile of sediments which made up the Southern Uplands Terrane was formed as a series of prisms of sediment, which were thrust against the edge of Laurentia by the advance of Iapetus ocean floor from the south-east.

Avalonia - Proterozoic

To the south, meanwhile, a quite different land mass was experiencing a different history. England, Wales, and the south-easterly two-thirds of Ireland, formed parts of the crustal block known as Avalonia, located along the edge of Gondwana. The Channel Islands formed part of the land mass termed Armorica, which is also thought to have been on the margin of Gondwana.

The Channel Islands and the adjacent parts of north-west France form part of a late Neoproterozoic orogenic belt, the Cadomian Belt. This is dominated by basic intrusions and volcanic sequences and by the deformed and metamorphosed Brioverian Supergroup, which is a sequence of sediments and volcanic rocks. Gneisses exposed on Guernsey have been measured at about 2 billion years old. The Cadomian Belt was affected subsequently by igneous intrusions about 600 million years ago. Subsequent rifting and sedimentation of the Brioverian supergroup were followed by continuing deformation and metamorphism to about 540 million years. Quartz diorites were emplaced in Guernsey and Sark during the first of these phases.

To the north, Avalonia is represented by two great Precambrian terranes, the Avalon Terrane which underlies central England and much of Wales; and the Monian-Rosslare Terrane of north-west Wales and south-east Ireland. The junction between these terranes passes through the Menai Strait.

Most of the Avalon Terrane is concealed by Cambrian and later rocks, but Neoproterozoic rocks occur as inliers, the alignment and position of which is controlled to a great extent by a series of long-lived deep seated crustal fractures, which include the Malvern Line or lineament, and the Welsh Borderlands Fault System. These rocks have been dated to the period from 700 to about 560 million years ago. Only in eastern Shropshire is any area of regional metamorphic rocks exposed. The small areas of exposure include a number of basic to acid igneous complexes together with some of their volcanic equivalents.

Only the sketchiest picture of the palaeogeography of the time can be attempted. The Uriconian group of Shropshire includes a thick pile of eruptive rocks from basic to acid composition, with indication of subaerial eruption. Somewhat similar rocks are exposed in parts of the Malvern Hills. Other volcanic and sedimentary sequences are exposed as the Arfon Group of north-west Wales, the Pebidian supergroup of Pembrokeshire (with a tiny, isolated inlier near Carmarthen) and the largely Longmyndian supergroup of Shropshire, which is calculated to comprise over 6 kilometres of mainly sandstones and mudstones. The rocks form a shallowing-upwards sequence and were deposited relatively rapidly, just before the onset of the Cambrian (some argue that the topmost Longmyndian is indeed Cambrian in age).

Near Leicester, the Charnian Supergroup comprises about 3 kilometres of volcanic and clastic sediments (noted for some of the earliest fossils, from the late Precambrian). Latest dating suggests that the topmost part of this sequence is, again, Cambrian.

The nature of the concealed part of the Avalon Terrane is largely unknown, since few boreholes have penetrated the basement.

The Monian-Rosslare Terrane has a complex history which recent improved dating techniques have failed to clarify satisfactorily. However, what has proved to be Precambrian include high grade gneisses, a granite intrusion, and a belt of metamorphic schists, with thick metasedimentary sequences of low grade.

The Rosslare complex of south-eastern Ireland corresponds with the rocks of Anglesey.

The current synthesis of available data suggests that the oldest rocks represent volcanic arc magmatism before about 680 million years and the main magmatic event about 600 million years ago. Subsequently, until about 550 million years, plate convergence caused the relevant land masses to form a complex series of small basins, in some of which thick sediments were deposited.

Cambrian to early Ordovician

The main stratigraphic divisions of the Cambrian are named in Table 5. Ordovician divisions are named in Table 6.

By 543 million years ago the rocks south of the present Solway line formed part of the northern or north-western margin of the major continental mass of Gondwana. It was realised by early geologists that there are major differences between Britain south of the Southern Uplands and the remainder, but no mechanism could be adduced to account for this. The differences in rock sequences, dates, and fossil faunas are readily accounted for by the concept of the Iapetus Ocean, referred to previously. Trilobites swam in the sea, together with unfamiliar arthropods, and most of the forerunners of today's fauna appeared in more or less recognisable form. By the end of the Cambrian, the earliest known land arthropods were leaving footmarks.

Table 5

Period/System	Age/Stage		Notes
Ordovician	Tremadoc		Classified as Cambrian on older maps
Cambrian 543-495 million years	Late Cambrian (Merioneth)	Divisions not easy to correlate outside Anglo-Welsh area. International usage is different	Includes most of Cambro-Ordovician of North-West Scotland
	Mid Cambrian (St David's)		
	Early Cambrian (Comley)		Appearance of significant macrofossil remains
Pre-Cambrian - Neoproterozoic	Ediacara		Rocks of the Long Mynd, Charnwood Forest; macrofossil remains rare and of unfamiliar types (Ediacara fauna)

Table 6

Period/System	Age/Stage	Notes
Ordovician 495-443 million years	Ashgill Caradoc Llanvirn Arenig	Much vulcanism in British areas; Grampian orogeny
	Tremadoc	Classified as Cambrian on older maps

Scotland and north-western Ireland were, at the beginning of the Cambrian, between approximately 40 and 25 degrees south latitude, on the southern edge of Laurentia, while England, Wales and south-eastern Ireland lay at about 60 degrees south, across a broad arm of the Iapetus Ocean. To the north or north-east was another significant landmass, Baltica, which is now represented by much of Norway, Sweden, Denmark and northern Germany. The margin of Gondwana, represented by Avalonia, had subsided and/or eroded after the preceding orogenic movements, to the point where marine transgression advanced from the north-west, probably aided by a major rise in sea level. The margin was passive during Cambrian time, with a number of basins developing, with deep marine sediments including turbidites accumulating in parts of Leinster, and in Wales (Figs.81 and 82). To the south-east of this, only shallow seas were present. In the extreme south of England, the history has been obscured by the development of the Variscides (see below).

By early Ordovician times, the continental margin had become active, although a number of basins were still filling with turbidites, including the Lake District Basin. At some point in the early Ordovician, Avalonia separated

Fig.81 The British part of Avalonia in mid-Cambrian times. Land areas tinted.

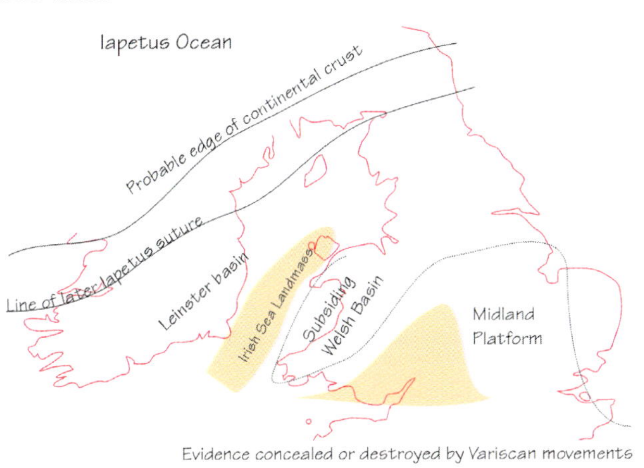

Fig.82 Upper Cambrian sandstone, Newgale, Pembs: bedding dip to right, prominent joint system dipping to left.

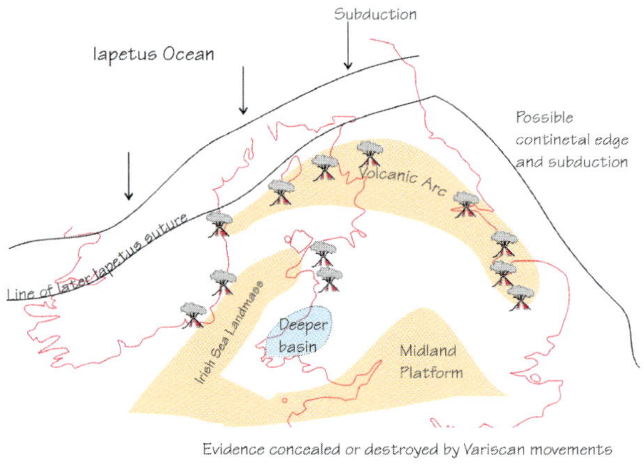

Fig.83 (left) The British part of Avalonia in upper Ordovician time. Land areas tinted. Fig.84 (right) Laminated Upper Ordovician mudrocks, nr Newport, Pembs.

from Gondwana, opening behind it the Rheic Ocean.

In front of Avalonia, which was proceeding to the north or north-west, the Iapetus Ocean began to narrow. By the earliest Silurian, Avalonia would begin to make contact with Laurentia. Ordovician rocks are exposed to a rather greater degree than Cambrian and Precambrian rocks, with extensive outcrops in south-eastern Ireland, and a few north of Dublin; they comprise much of the Isle of Man. There is a large outcrop in the centre of the Lake District, extensive outcrops in North and South Wales, and smaller but still significant outcrops in the Welsh borderland. Otherwise, most rocks of this period are concealed or missing.

The period is characterised by much more volcanic activity and a volcanic arc is thought to have been present in at least two places, running through Leinster into the Isle of Man and through the Welsh Borderlands respectively. Thick clastic sedimentation, often with volcanic inclusions, accumulated in much of Wales (Figs.83 and 84) and in a basin which extended from Leinster to the Lake District, respectively. The earliest Ordovician rocks are about 440 million years old, but the palaeogeography can be constructed with much more confidence than for earlier times.

It is thought that the Iapetus Ocean crust was subducting to the south-east under Avalonia, giving rise to the formation of the volcanic arcs, in the same way as the Indian Ocean plate subducts under Indonesia today. The volcanic climax occurred in the Mid- to Late Ordovician, and some of the land mass emerged above sea level at this time.

When reading maps, bear in mind that until quite recently the Geological Survey treated the lowest Ordovician (The Tremadoc Series) as part of the Cambrian.

The Silurian

Table 7 shows the main Silurian Divisions.

During Silurian times, convergence with Laurentia, at an oblique angle, seems to have continued. The Silurian period ranges from about 443 to 410 million years ago. Silurian rocks are exposed extensively throughout Wales, the Lake District, and various parts of south-eastern Ireland. There are smaller but still important outcrops in the Mendip Hills and in the Tortworth area, north of Bristol.

The Silurian successions are predominantly sedimentary without great accumulations of volcanics. This is thought to reflect the shutting down of the volcanic arcs in the Lake

Table 7

Period/System	Age/Stage	Notes
Silurian 443-418 million years	Prídolí	Onset of Old Red Sandstone facies
	Ludlow	Includes limestones of the classic Welsh Borders and Dudley areas; end of British graptolites, used for dark mudrock correlation
	Wenlock	
	Llandovery	

District, Wales and Leinster, related to subduction zones. The successions in general show an essentially conformable upward transition to deltaic and riverine sediments, generally by the end of the Silurian. Deep basins were present in Leinster, the Lake District and Wales, but shallower water was present elsewhere (Fig.85), and in some cases there is a break between the Ordovician and the lowest Silurian (Fig.86). In this shallower sea were deposited the well-known and often highly fossiliferous limestones of the Welsh Borderlands. In the basins, however, great thicknesses of turbidites and shales accumulated.

Towards the end of the period, graptolites, marine colonial organisms which are useful for correlation in dark shale facies, died out in the British area. Trilobites and brachiopods are among the commonest macrofossils, and the earliest fish appeared.

Beginning in early Wenlock times, the edge of the Avalonian landmass was progressively overridden by the edge of Laurentia, approaching at an angle from the north. This process, the Acadian Orogeny (part of the Caledonian Orogeny), had probably finished by late Ludlow to Prídolí times, by which time mountain ranges raised by the orogeny were feeding reddish sediments, from subaerial erosion, into the remains of the existing basins. This reddish sediment forms the onset of the so-called Old Red Sandstone, which at first was thought to be only of Devonian age; it has, since the 1970s, been realised that the lowest section in some areas is topmost Silurian in age. By the end of the Silurian, Laurentia and Avalonia were locked together, with only the southern part of the present British area to be added.

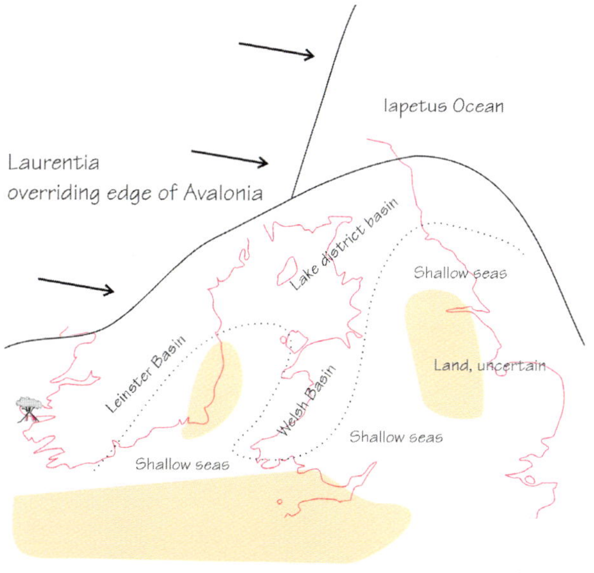

Fig.85 (top) The British part of Avalonia in mid-Silurian times. Land areas tinted. Fig.86 (above) Lower Silurian (Llandovery) resting with minor but distinct angular unconformity (indicated) on late Ordovician (Caradoc), Shropshire.

The Caledonian Orogeny

The Caledonian Orogen is still the major visible structural feature of the British Isles. A glance at the geological map (Fig.79) shows that many of the outcrops in parts of Ireland, Wales, the Lake District and Scotland are aligned north-east to south–west.

This trend, the 'Caledonoid' trend, reflects the structures which were imposed on the united land masses of Laurentia and Avalonia by the orogeny. The outer edges of the orogen are conventionally taken as follows. In north-western Scotland, the Moine Thrust, the sinuous outcrop of which roughly parallels the metamorphosed north-western seaboard, marks the thrusting of the metamorphosed Scottish Highland rocks over the Lewisian gneiss Hebridean platform.

To the south-east, the Welsh Borderland Fault System marks the boundary between the strongly folded and cleaved rocks of the Welsh basin which are juxtaposed and up-faulted against the less strongly deformed Midland platform. This deformation front has been traced in the subsurface, below later rocks, around to south-eastern England, but so far little is known otherwise. The southern Caledonian deformation front is obscured and confused at both ends by the Variscan Front (explained below).

The orogeny is thought to have consolidated most of the British and Irish crust into its present layout, with the sequence of mostly fault-bounded terranes indicated on the map (Fig.80).

The mountains created by the orogeny were the prime source of sedimentation for the surrounding Old Red

Sandstone basins of late Silurian and Devonian age.

The orogen was also included extensively during Silurian and Devonian times, by mainly granitic inclusions, over a period of some 60 million years.

The Old Red Sandstone Continent

Table 8 lists the principal divisions of the Devonian.

Laurentia and Avalonia together (and Baltica to the east, which we have not discussed) is termed Laurussia. However it has also been termed 'the Old Red Sandstone Continent' because of the non-marine, predominantly reddish, sedimentary rocks preserved over much of the continent.

Table 8

Period/System	Age/Stage	Notes
Devonian 418-362 million years	**Famennian Frasnian Givetian Eifelian Emsian Pragian Lochkovian**	**Continental sedimentation (Old Red Sandstone) throughout most of British area; marine deposits in Devon and Cornwall; peak of Caledonian Orogeny**

At the time, Gondwana lay somewhere, probably fairly close by to the south (see below). The sea level was at its lowest near the Silurian-Devonian boundary and gradually rose to its highest point within the late Devonian, falling and rising again before the onset of the Carboniferous.

It seems that the Devonian Old Red Sandstone rocks (Figs.86, 87 and 44) were deposited in a number of largely separate basins, which were partly defined by the recently created mountain ranges and fault-bounded basin margins.

The Devonian system is named after Devon, where it was originally defined by the early 19th-century pioneers Adam Sedgwick and Roderick Murchison. Unfortunately, the rocks in Devon are so complex that they are unsuitable as an international type section. Nevertheless, they can be correlated, at least in part, with the reference sections of continental Europe. However, the non-marine successions in the basins to the north are not so readily correlated due to their localised and sporadic faunas, mainly concentrated in so-called 'fish bands'. Consequently, the early terms, Lower, Middle and Upper Old Red Sandstone are still widely used for convenience, although these equate crudely or not at all with the faunally described chronostratigraphic series. As noted, the Old Red Sandstone facies generally began to appear in the Prídolí Series (named after a Czech locality), but in some areas, such as extreme south-west Wales, south-west Ireland and parts of eastern Scotland, this change probably occurs during

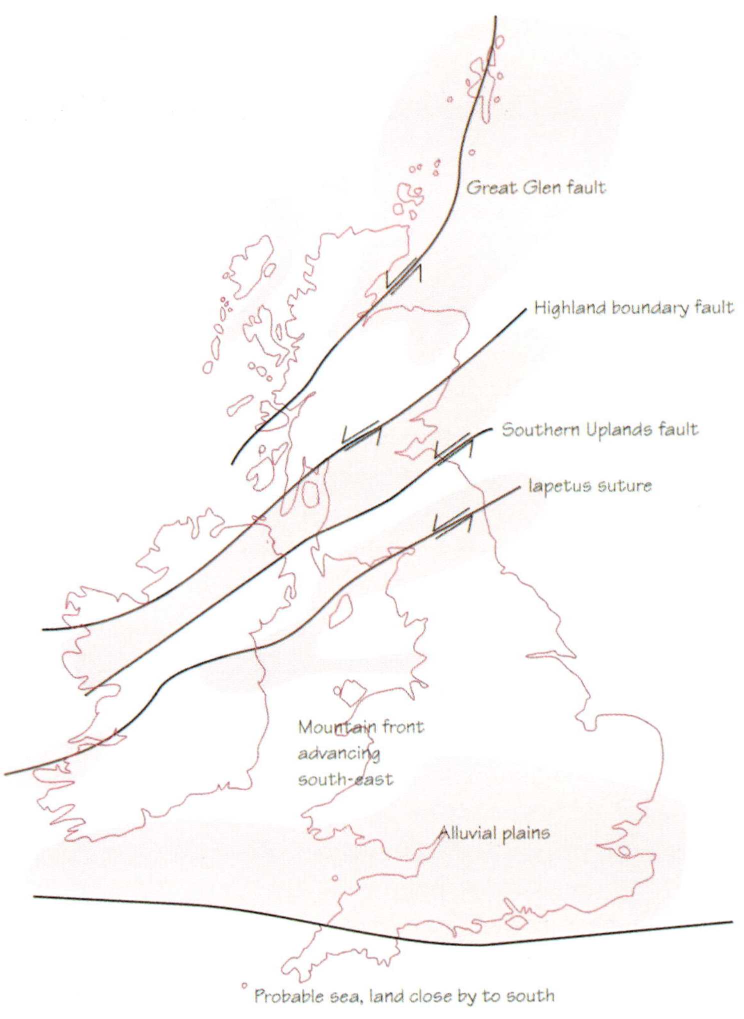

Fig.87 The British area in lower Devonian times. Laurentia (Scotland and north-west Ireland) have joined Avalonia. Depositional basins tinted.

Fig.88 Somewhere in this view is the boundary between the Devonian and Carboniferous. Note the colour change from red to grey, which at this location corresponds closely to the stratigraphic boundary. Milford Haven.

the earlier Ludlow. In the same way, the topmost Old Red Sandstone extends in some areas into the Lower Carboniferous (Fig.88).

The Old Red Sandstone in general comprises sandstones and mudstones of continental aspect, including obvious alluvial fans, flood plain deposits, etc, several of which include fossil soil horizons with calcareous concretions (calcretes), known in Britain as 'cornstones' (Fig.44). These often form resistant topographic features.

In northernmost Scotland and the Orkneys, middle Devonian rocks are present, but these are absent in the Anglo-Welsh basin, suggesting uplift due to the Arcadian Orogeny. The basins of Scotland, however, were internal to the mountain massifs and continued to receive sediment. Rising sea level drove marine transgressions onto the edge of the Anglo-Welsh shelf, while there was also some marine influence penetrating some of the Scottish basins. These basins include the flaggy sandstone deposits of Caithness and Orkney.

To the south, open sea lay along the southern margin of Avalonia. The precise relationship with the Avalonian land mass is now obscured, because of the deformation during the Variscan Orogeny. It is now thought, however, that during Devonian time, the Devonian rocks of present-day Devon and Cornwall may well have lain further to the east than their present position to the south of Wales. Along the line of the Bristol Channel was a zone of occasionally emergent Precambrian basement with, to its south, the Exmoor basin with a mixture of marine and some fluvial sediments. The transition to deeper water deposits to the south is now obscured by the overlying Carboniferous Culm Basin (Fig.89).

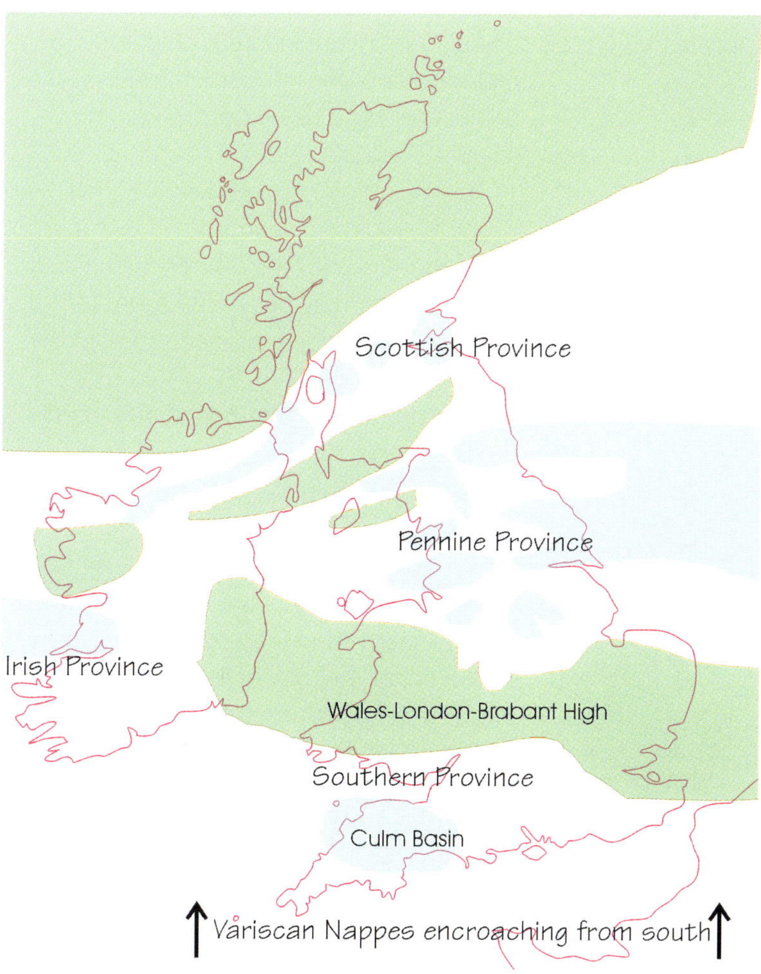

Fig.89 The British Isles area during Lower Carboniferous times. Land areas tinted green; deeper basins tinted blue.

To the south of the Culm Basin, the Devonian rocks show features suggestive of a northerly, fault-bounded margin to deep marine basins to the south. Yet further to the south there are indications that sediment had been supplied from landmasses near present-day Brittany. There are strong indications of the presence of oceanic as opposed to continental crust, including the rocks of the Lizard. These include suspected ocean floor remnants, which have been thrust over the Devonian rock. Shallower areas in the Devonian seaway supported limestone growth including those of the Torquay area.

By the end of the period, the landmass of Armorica (basically, Brittany) to the south was almost certainly already in contact with the Avalonian continental crust and the Devonian basins were already being deformed and squeezed by the earliest phases of the Variscan orogeny.

During the Devonian, the first, rather fish-like amphibians began to become a significant part of the fauna. Land plants appeared, and with them the insects that ate them (some of the earliest of both are preserved in cherts at Rhynie, Scotland).

The Carboniferous Period

The main divisions of the British Carboniferous are set out in Table 9.

Table 9

Period/System	Epoch/Series	Age/Stage	Notes	
Carboniferous 362 -290 million years	Silesian	Stephanian		"Coal measures" with many coals; reddened beds towards top with no coals
		Westphalian	Culm Basin deposits of Devon and Cornwall	
		Namurian		Much deltaic sandstone; few coals
	Dinantian	Visean		Mostly carbonate formation; coals in Scotland
		Tournaisian		

During the Carboniferous, rock groups were deposited which include some of the most conspicuous and economically important deposits of the British Isles, namely the limestones of the Lower Carboniferous and the coal-bearing sediments of the mainly Middle and Upper Carboniferous. Commercial extraction of the limestones, coals and ironstones of these deposits have provided very large quantities of information compared with the less tractable or economically less valuable rocks of other periods. The regional stratigraphy of the British Isles had been well established by the early 20th century, although it is still being refined and late 20th-century oil and gas discoveries in Carboniferous rocks below the North Sea have provided new sources of data.

The two major successions are the Lower Carboniferous, or Dinantian (Fig.89), and the Upper Carboniferous or Silesian (Fig.90). The boundary may be revised soon, but this will be of little practical significance to the majority of readers.

By the onset of the Dinantian, a northward-dipping subduction zone was in existence across central France and southern Germany. North of this, and behind a volcanic arc, was a seaway floored by oceanic crust, which ran across the present south-west England into northern France and Germany. Further to the north, the rest of Britain was divided into a series of basins by three long-lived, persistent landmasses: the Highlands, the Southern Uplands High (with extension into Ireland) and the Wales-London-Brabant High including an extension into the Leinster

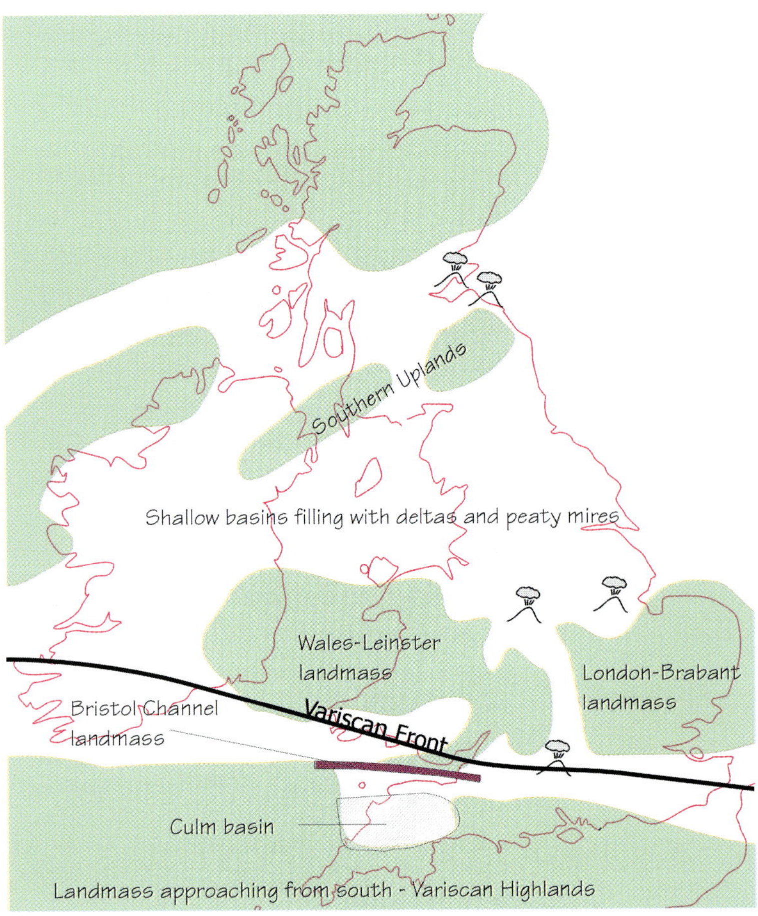

Fig.90 The British Isles area during Upper Carboniferous times: land areas tinted.

Massif of south-eastern Ireland. The degree of connection between the relevant basins varied. The most complete Carboniferous sequences were found in the area now occupied by the Pennines and it is here that most work has been done.

During the Dinantian, to the north of the Wales-London-Brabant High there was a complex of extensional rifts or basins. To the north of this, deposition was dominated by siliciclastic sediments with carbonate sedimentation more important in the south. In the Scottish province, a series of small, linked basins accepted a series of sediments formed under lacustrine, marginal marine and fluvial deltaic conditions which were also significantly influenced by volcanic activity and faulting. Volcanic eruptions were followed by the intrusion of numerous sills (some of which converted adjacent coal seams to anthracite or even the equivalent of coke).

Fig.91 Mudrocks and thin sandstones pass up to a thin coal and carbonaceous shale, overlain by massive sandstones. Upper Carboniferous (Silesian) of Pembrokeshire.

South of the Wales-London-Brabant-High, throughout the Dinantian, there was a southward-sloping carbonate 'ramp'. Further still to the south, the Culm Basin accumulated a series of deep water, fine grained sediments followed later by siliciclastic sediments including abundant turbidites.

In the Pennines and Ireland, carbonate deposition was continuous into the lowermost Upper Carboniferous. However the Silesian was otherwise dominated by siliciclastic sedimentation, while in northern England, several major deltaic systems had deposited extensive, thick, sand bodies by the early Upper Carboniferous.

The Westphalian was dominated by non-marine sedimentation including coal-forming mires (Fig.91), although deep-water conditions were still apparently present in the Culm Basin. In the Early Westphalian, there were several marine incursions, which gave rise to the so-called 'marine bands' which have been much used for correlation. Thick mire deposit accumulated in a humid or tropical climate on a low deltaic plain that extended across much of Western Europe.

During the Carboniferous, the continued movement of Armorica gave rise to the Variscan Orogeny, with a mountain range running roughly east-west across Cornwall and Devon. These rising mountains caused great contortion of the Culm Basin (Fig.92) and shed coarse sediment into south Wales and southern England (forming the massive Pennant Sandstones of Bristol and South Wales (Fig.93, see overleaf), much used for building). Some of this sediment was transported over the Wales-London-Brabant High into the southern part of the Pennine Province.

Towards the very end of the Carboniferous, the humid conditions had already disappeared and a long arid period had set in, which would endure during most of the following Permian and Triassic periods. This aridity is reflected in the presence of so-called 'red beds' in the uppermost Coal Measures. These red beds are barren of coal seams. The topmost Carboniferous is poorly represented in Britain and Ireland

Fig.92 Results of the Variscan Orogeny: lower Carboniferous black slates thrust over pale Devonian slates. Tintagel, Cornwall.

Fig.93 Cliff of Upper Carboniferous sandstone (Pennant sandstone), South Wales.

because of the Variscan uplift; late Carboniferous igneous activity produced dolerite dykes and sills in Scotland, Ireland and northern England (including the massive Clyde volcanic plateau and the Whin Sill). To the south, the granites of Dartmoor and Cornwall were emplaced and eroded pebbles from these were probably already accumulating in sediment by the very end of the Carboniferous.

The so-called Variscan Front (Fig.80), which marks the northern boundary of the more intensive Variscan movement, follows a fairly well defined line across the corner of south-western Ireland, the edge of south Wales, north of Bristol, and then under the later sediment cover in south-eastern England. The Lizard complex of Cornwall is thought to be a remnant of ocean floor thrust upwards onto a section of continental crust at some time during the Variscan Orogeny. The final emplacement of the Lizard rocks occurred after about 360 million years ago, since late Devonian fossils were found in rocks immediately beneath the bounding fault. The complexity of structures in mid-Devon and Cornwall is of Alpine proportions and only recently has it proved possible to understand how some of the rock masses are related.

The stress and temperatures of the orogeny produced relatively low-grade metamorphism throughout most of Cornwall and southern Devon, although high-grade metamorphic rocks are rare. The metamorphism released mineral-laden fluids, which gave rise to the numerous quartz-carbonate veins, which carry so much of the mineral wealth of this area.

To the north of the Variscan Front, a number of older structural fractures were reactivated in various ways, with predominant dextral strike-slip movements. This includes the Highland Boundary and Southern Uplands Faults, and the roughly north-south series of structures which runs from the Malvern Line north to the western border of the Pennines.

Amphibia continued to be an important element in the vertebrate fauna, and by the end of the period the first fully terrestrial quadrupeds were well established. The atmosphere may have had a higher oxygen content than today, permitting the appearance of huge dragonflies.

The Permian to Triassic Desert

Even today the Permo-Triassic stratigraphy (Tables 10 and 11) is problematical. The paucity of fossils, combined with a lack of easily correlatable horizons such as widespread tuffs, led to a somewhat arbitrary early division of the Permo-Trias and only recently has it been realised that some areas previously recorded as Lower Triassic are in fact Permian, while some of the so-called Permian probably include small amounts of Upper Carboniferous rock. Much of the rock is red or reddish, giving rise to the old term 'New Red Sandstone', which is now largely disused although you may come across it in literature.

Following the Variscan Orogeny, there was a youthful, high-relief topography of faulted mountain masses and exposed

Table 10

Period/System	Age/Stage	Notes
Permian 290-248 million years	**Zechstein**	**Evaporite sea stretching into German area from northern Britain; "Magnesian Limestone"; major faunal extinction at end**
	Rotliegendes	**Mostly desert sandstones, etc formed by erosion of Variscan mountains**

volcanic rocks, which were being actively eroded in a dry climate. Post-orogenic subsidence of some areas began, notably in the western approaches and in the North Sea. The first rifting, a forerunner of the eventual formation of the North Atlantic, began. The area was largely desert or semi-desert, characterised by flash floods, isolated lacustrine deposits, and desert dune deposits. However, during the Permian there was a rise in sea level which caused the so-called Zechstein Transgression, named after a German locality, which gave rise to limestone and evaporite deposits that now form the limestone and evaporite belt running from Nottingham northwards to Teesside. These represent only the edge of a much larger evaporite basin stretching into Germany (Fig.94).

Further evaporite deposition occurred during the Trias, including the salt deposits of Cheshire and Worcestershire. At the very end of the Triassic, by which time the topography had become largely subdued, the sea began to transgress again.

The Permian and especially Triassic sandstones now form important oil reservoirs (including the largest on-shore oil field in Europe, the Wytch Farm Field under Poole harbour).

The Variscan collision between Armorica and Avalonia was part of a process which ended in the construction of the huge supercontinent of Pangaea. This continent persisted during Permian and Triassic times, while rotating so that Britain and Ireland moved from a Carboniferous equatorial position to something like 25 degrees north by the late Permian and as far as 40 degrees north by the Jurassic. During this long time span the continent was affected by rift zones which would ultimately become lines of break-up, leading to the formation of land masses which are recognisable, at least to some degree, as the modern continents.

Among the sparse fossils from this period, early reptiles have been found in some British strata.

The end of the Permian is notable for the so-called 'mass extinction' during which life forms were apparently reduced both in number and in diversity. Scientific argument about the causes of this and other mass extinctions remain fierce, but the Permian is arguably the most significant one. At this time, whole groups of animals died out for ever, including the Trilobites,

Table 11

Period/System	Age/Stage	Notes
Triassic 248-206 million years	**Rhaetian Norian Carnian Ladinian Anisian Scythian**	**Mostly desert and lake deposits; salt deposits of Cheshire, Worcestershire**

Fig.94 Generalised Permian palaeogeography of the British area: basin areas tinted.

certain types of coral and less well known groups, while those groups which persisted often show a marked change in character. The Brachiopods, the bottom-dwelling filter feeders characteristic of the Palaeozoic, were greatly reduced in number and have dwindled to very limited numbers today.

Many authorities consider that the Carboniferous limestone hills of extreme South Wales and the Bristol/Mendip area are an 'exhumed landscape' that resembles, if only broadly, the late Triassic topography before the sea advanced again in the early Jurassic. In fissures in the limestone, the bones of tiny early mammals are sometimes found within the Triassic sediments.

Jurassic

Table 12 sets out the divisions of the Jurassic.

By the end of the Jurassic, the supercontinent of Pangaea was a more or less U-shaped mass opening to the east in a broad, tropical seaway known as Tethys. The remaining part of the world ocean was called Panthalassa. The continental distribution is thought to have given rise to a pattern of climates which was recognisably zoned latitudinally like that of the present day.

In north-west Europe generally, the structural grain built into the land mass by previous earth movement exercised considerable influence over the development of new sedimentary basins. For instance, the Cardigan Bay Basin (mostly underwater) and the Cheshire Basin are approximately oval, of similar dimensions, and orientated north-east to south-west; while many structures in southern England are parallel to the earlier Variscan trend lines. The record of Jurassic and later rocks in Ireland is particularly sparse and most of the history of the Irish area can only be inferred from the scanty deposits found in Ireland and the evidence from the surrounding seas and on the Great Britain landmass.

At the end of the Triassic, there was a major transgression, caused partly by global sea level rise and partly by earth movements. Marine deposition ensued over much of the British area, and resulted in a local connection between the Tethys and an ocean to the north, which one might term 'the Arctic'. The dry climate of the Triassic changed to humid conditions, and the transgression caused widespread deposition of mudrocks. Many of these have a very high organic content, which has subsequently formed the basis for petroleum deposits. This pattern of regional minor movements and regression resulted in formation of shallow marine sands. In the mid-Jurassic, sedimentary deposits were largely of regressive nature, with shallow-water carbonate ramps, and to the north, deltas and coastal plains (Fig.95). The deltas included the formation of carbonaceous shales, coaly streaks and (at Brora, in northern Scotland) a workable coal seam. During this time, there was widespread erosion of pre-existing strata. In the central North Sea, there was significant igneous activity.

By the late Jurassic, crustal extension and the formation of rifts, especially in the area of the

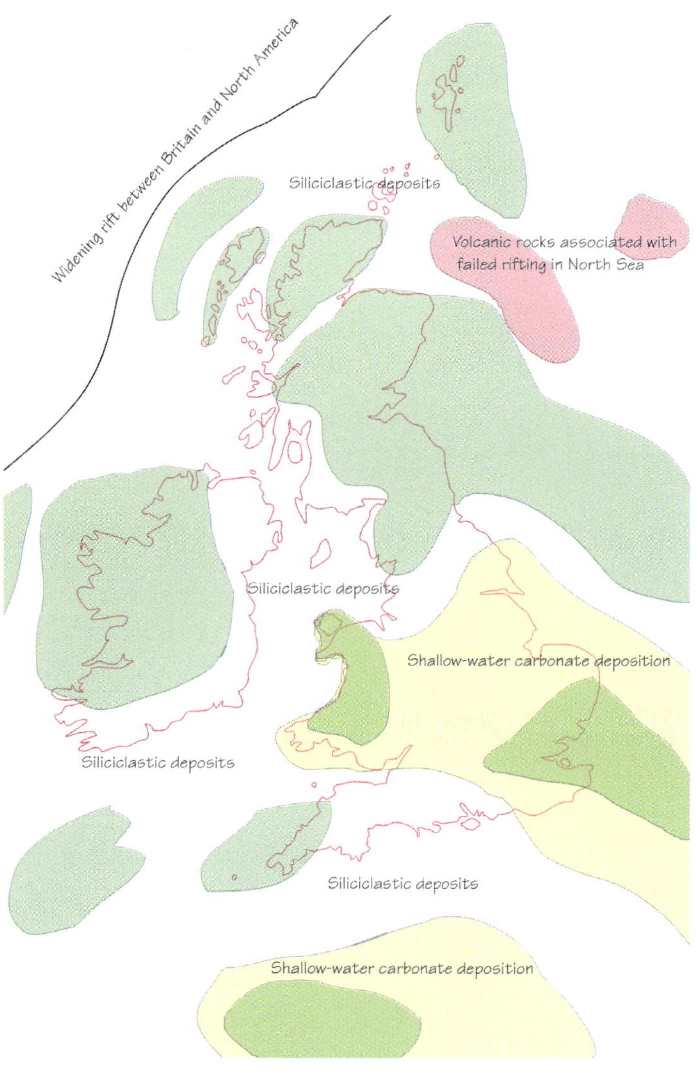

Fig.95 Mid-Jurassic palaeogeography of the British Isles. Land tinted green.

northern North Sea were the first indication of the potential separation by modern North America. Sediment supply in northerly Britain and Ireland became limited, while abundant sediments continued to be formed in many southern provinces. The Upper Jurassic in southern Britain comprises mainly mudrocks, but with important and persistent carbonate horizons. There are signs of some localised aridity, at least at the end of the Jurassic, resulting in formation in minor evaporite deposits, including the gypsum extracted by British Gypsum in Sussex. Rifting ceased by the end of the Jurassic, but continued and progressed on the Atlantic margins of the present British area.

Table 12

Period/System	Age/Stage	Notes
Jurassic 206-144 million years	Tithonian	Purbeck evaporites
	Kimmeridgian Oxfordian Callovian	Oxford and Kimmeridge clay formations,much used for brickmaking
	Bathonian	Important oölitic building stones
	Bajocian Aalenian Toarcian	
	Pliensbachian Sinemurian Hettangian	Rise of dinosaurs; Lias clays and mudstones

Cretaceous

Table 13 sets out the divisions of the Cretaceous.

Most of the Cretaceous was characterised by a 'greenhouse' climate, apparently with only a weak thermal gradient from the equator to the poles and no signs of polar ice at low altitudes. In the Mid-Cretaceous generally, there was a great increase in submarine volcanism which resulted in increased ocean ridge production in the Pacific. One of the results of this was a very high sea level in the later Cretaceous.

In the early part of the Upper Cretaceous, the old Palaeozoic massifs, which had exercised significant control on the sediment pattern of the Jurassic, continued to be significant. Many of these were probably partially or completely submerged (Fig.96, see overleaf) by the end of the Lower Cretaceous.

In Southern England there is a sudden transition from the limestones and shales of the topmost Jurassic, with evaporite formation, to the clastic facies of the Wealden Group. A long history of boring, mainly for oil, in the Wealden and Wessex areas has allowed reasonable reconstruction of the Wealden palaeoenvironment. There is a lower Hastings Group and a muddier upper Weald Clay Group. Freshwater floodplains and braidplains (a plain formed by the deposition of braided rivers) were formed, with only sporadic marine incursions.

Towards the end of the Lower Cretaceous, the marine transgression deposited the Lower Greensand. This is, notoriously, often neither green nor sand. The Greensand term comes from the incorporation of glauconite, a greenish clay mineral, which once exposed rapidly oxidises to browner colours.

A further major sea level rise, which extended into the Upper Cretaceous, resulted in the deposits of the thin but very persistent Gault Clay, which gives way in more westerly regions to sandy deposits near the former shore lines. Above this, the general spread of ocean waters onto the

Table 13

Period/System	Epoch/Series	Age/Stage	Notes
Cretaceous 144-65 million years	Late or Upper	Maastrichtian Campanian Santonian Coniacian Turonian Cenomanian	Deposition of Chalk
	Early or Lower	Albian	Gault and Upper Greensand marine transgression
		Aptian Barremian Hauterivian Valanginian Berriasian	Wealden and Wessex basin deposits; sparse further north

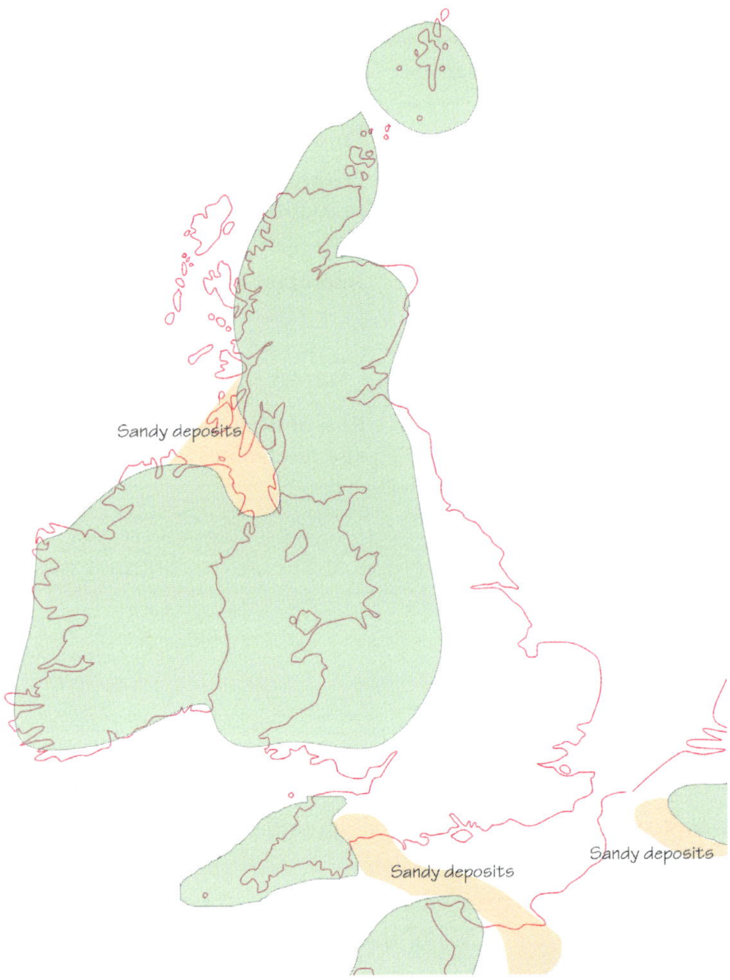

Fig.96 The British Isles area during the earliest Upper Cretaceous. Land tinted green.

land mass margins resulted in greatly reduced terrigenous input to the sea, and the resulting deposition of the Chalk, which shows a progressive diminution in clastic components upwards.

In the east Midlands and the North Sea margin, thin and much-reworked deposits accumulated during Lower Cretaceous times and are again overlain by the Chalk. Detailed work by many scientists has enabled a reconstruction of the trangressions and regressions during the Cretaceous which is too voluminous to record here.

It is thought that during the Chalk deposition period arid conditions persisted on the surrounding exposed land. The sea level may, in relative terms, have been the highest in Cambrian and later times. Chalk deposition continued through the entirety of the Upper Cretaceous. Included flint beds often have a distinctive morphology and recent work has shown that they can be traced over extremely large areas for correlation purposes. Chalk deposits extend as far north as the northern North Sea, where they give way northwards to marly Chalks or mudstones. The North Sea oil exploration has greatly increased knowledge of the Chalk generally. In the North Sea and in Denmark, Chalk continued to be deposited past the end of the Cretaceous into the earliest Palaeocene. In England, there is extensive folding upwards and erosion of Chalk, resulting in a dramatic facies change to the continental, terrigenous sedimentation of the Tertiary. Regionally, folding was generally very gentle. On mainland Britain, most of the very latest Chalk has been eroded although some is preserved offshore. Very late Chalk is preserved in north-east Norfolk, where great slabs of Chalk displaced by glacial action have been pushed up onto the present land.

The Tertiary

The Tertiary contains a number of divisions that were formerly ranked as Periods/Systems but have been demoted (Table 14) so that all periods of the Phanerozoic are of roughly equivalent length (apart from the Quaternary, which includes the present day).

The stratigraphic framework of the Tertiary is now extremely well known as a result of much research. Ocean-bottom sediment cores have recovered a near-continuous record of Tertiary sedimentation. This has been analysed for lithology, biology, isotopic signatures and palaeomagnetic signatures. Key dates in the system have been provided by radiometric dating of volcanic horizons.

The boundary between the Cretaceous and Tertiary is significant worldwide and is popularly known to scientists as the 'K/T Boundary'. It is famous for the controversy surrounding the disappearance not only of the dinosaurs, but of many other taxonomic groups. This has been associated by many workers (although controversy over the details is strong) with the impact of a meteoroid or asteroid which may have struck part of Mexico. Although such an event could have had a devastating effect, it was not necessarily the critical factor, although it must have

Table 14

Period/System	Sub-Period/Subsystem	Age/Stage	Notes	
Quaternary 1.8-0 million years		Holocene	Now. Probably an interglacial	
		Pleistocene	Glacial and interglacial periods	
Tertiary 65-1.8 million years	Neogene	Pliocene	Sparse deposits	
		Miocene	No known British deposits	
	Palaeogene	Oligocene		Deposits on land mostly confined to London and Hampshire basins, mostly terrestrial with some marine incursions
		Eocene	Volcanoes and lavas of Scotland and Antrim	
		Palaeocene		

contributed. However, in Britain, the K/T Boundary is characterised by an unconformity between the Cretaceous Chalk and the Tertiary.

During the Late Cretaceous, sea floor spreading and the growth of the Atlantic Ocean had progressed only as far north as the Labrador Sea. During the earliest Tertiary, the Palaeocene, a major focus of volcanic activity developed in the area between the present Faeroe Isles and Greenland, which caused thermal uplift and massive vulcanism over a very large area including parts of western Scotland and Northern Ireland. At about the beginning of the Eocene, rifting and continental rupture developed across this thermal bulge and the northern part of the north Atlantic began to form. From Eocene time onwards, the British area formed part of the general north-western European continental margin, a so-called 'passive margin', which by ocean floor spreading moved progressively away from the volcanic focus. This 'hot spot' appears to continue at the present day under the volcanically active Iceland.

Meanwhile, to the south, Africa was colliding with Europe and closing the remnants of the Tethys Ocean, referred to above. This resulted in the Alpine orogeny, which gave rise to the Alps and the Pyrenees along a complex collisional plate boundary. In north-west Britain and Ireland, much of the crust was lifted above sea level and subaerially eroded. Considerable thicknesses of basaltic lava were erupted onto this surface and substantial remnants of this remain both on the continental shelf and on shore. The lavas are developed especially in Skye, Mull, and Country Antrim, and in the Hebrides they are up to 2 kilometres thick (towards the north-west they increase to 5 kilometres on the Faeroe Islands). Thin sediments, the natures of which suggest a humid, temperate climate, are found between individual lava flows. Lavas were mostly erupted from fissures (now preserved as dykes) as the magmatism produced local central complexes of both acid and basic plutonic masses associated with a wide variety of sills, dykes and other hypabyssal intrusions. These intrusions cut, and therefore are later than, the lava fields. The most southerly manifestation of this igneous activity is the Tertiary granite of Lundy Island, in the Bristol Channel. The dyke 'swarms' associated with the volcanic centres are sometimes of great extent; for instance, the Acklington Dyke of northern England appears to radiate from a volcanic centre in Mull, Scotland.

Sediments in the offshore areas (Fig.97) include lignitic coals. Only small remnants of Tertiary sediment are found on shore in the northern and western British Isles, such as

Fig.97 The present onshore Tertiary basins are the tips of larger offshore basins (tinted).

the Lough Neagh Basin of Ulster. The Lough Neagh sediments include lignite beds tens of metres thick, formed from plant debris washed into a lake.

In the North Sea, the basins which had been starved of sediment during the Chalk deposition began to fill with clastic sediments reaching a thickness in the order of 3 kilometres at their deepest. Much of the included sands were clearly derived from the Scottish land mass. Again, coals (lignites) are quite common.

South-eastern England includes the only preserved areas of major Tertiary sediments. These date only from the late Palaeocene, the Eocene and the earliest Oligocene. Outliers are found quite far to the west on the Chalk. The hard Sarsen stones of which the prehistoric monuments of Stonehenge and Avebury are made are siliceous sandstone that probably represent the remains of early Tertiary 'hard pans' or 'duricrusts', fossil soils in which minerals are concentrated by leaching in tropical conditions.

Onshore thicknesses are 500 metres at best (in the Hampshire Basin), although in the southern North Sea the maximum thickness is up to 2 kilometres. The sediments are terrigenous and reflect repeated minor transgressions of the sea with alternating repetitive cycles of sands, mudrocks and sporadic limestones. The most notable Tertiary unit is the London Clay Formation, which is found throughout the London and Hampshire Basins. This is thought to have formed during the most widespread Tertiary transgressions, as only the marginal areas of the deposit are at all sandy. Included fossils suggest that the Tertiary climate was very warm, with the British area covered by dense tropical forest.

Pliocene and Pleistocene sediments are extremely rare, but the Lenham Beds, which are of Pliocene age, are found in solution pipes in the Kentish Chalk, while thin but relatively extensive Pliocene carbonate sands are found in Suffolk.

To the west, isolated basins were formed in Tremadoc Bay, at Bovey Tracey in Devon, and in a few other scattered localities. The Tremadoc Bay Basin was unsuspected until a borehole was drilled at Mochras near Harlech and proved the presence of 500 metres of Oligocene and Miocene rocks on Jurassic below. Here, coarse clastic deposits pass upwards into sands, silts, clays, and lignites, the evidence pointing to floodplains with occasional mires. The basins in Devon also contain Oligocene clays, sands and lignites, some of which are of great thickness. The Bovey Basin is only 6 kilometres wide, but contains over 1 kilometre of Eocene to Oligocene sediments. These basins have been preserved by movements associated with a major fault zone which traverses south-western Britain, the Sticklepath-Lustleigh Fault zone. Larger areas of Tertiary sediments are preserved off-shore and appear to be largely similar, but to the south, in the present English Channel, sediments record more persistent marine conditions throughout the tertiary. The several phases of uplift and erosion throughout the Tertiary can be equated to the tectonic events in the Alpine and Pyrenean Orogenies.

Cooling from Eocene times onwards was slow, and tropical plants and large mammals such as hippopotami populated Britain until quite recently.

The Quaternary

The boundary between the Pliocene and the Pleistocene, the lower part of the Quaternary, is defined in a series of marine sediments in southern Italy. This is dated at about 1.8 million years ago, but the Quaternary in north-western Europe, as defined by reference to climatic deterioration, has been dated at about 2.4 million years. This early date remains in informal use in British and Irish work.

The Quaternary is marked by persistent glaciations. Availability of cores and other samples from continuous marine sediments outside the British area shows, by oxygen isotope proportions, repeated cyclical changes in water temperature which can be correlated with the successive glacial advances. The onshore record in Britain, mostly in the Midlands and East Anglia, suggests eight major alternations of cold and more temperate climates. These are commonly described as 'glacial' and 'interglacial' periods respectively. Only the last three glacials, however (the Anglian, Wolstonian and Devensian) can be definitely correlated with onshore tills and other sediments. Fig.98 shows the Wolstonian and Devensian ice sheets at their maximum extents. At the onset of the Pleistocene, Britain would probably have been recognisable to a present day eye. However, the glacial intervals resulted in significant alteration of the topography including the formation of the glacial hollows (cirques, corries or cwms) of northern and western Britain and their associated glacial troughs.

Changes in sea level gave rise to the formation of beaches and shore platforms at levels that are now below or above sea level. Sometimes beach deposits, often of boulders, cobbles and gravel have survived as a result of rapid cementation; these are termed raised beaches (Fig.99).

The complex and discontinuous pattern of tills and associated sediments includes not only the deposits laid down under the glaciers as they advanced and melted, but also the sediments derived from the reworking of these earlier sediments and the muds and silts laid down in lakes trapped within the ice margin and high ground or formed in hollows made by the ice. Major river diversions occurred. For instance the headwaters of the River Severn are thought to have flowed originally to join the River Dee or its precursor. The River Thames formerly flowed much further to the north of the London area. It reached the North Sea in the latitude of Suffolk and was joined by an early River Medway, the river terraces of which are preserved in eastern Essex. In upland areas, many steep-sided channels were eroded by water escaping from glacially dammed lakes; many of these now feature tiny misfit stream beds.

Characteristic depositional features in Britain include various terrace-like deposits along river valleys resulting from the deposition of lateral glacial moraines; long, sinuous sandy deposits, eskers, which were deposited by streams within or emerging from glaciers; and drumlins, which form fields of gentle, elongated humps, like those which form islands in Strangford Lough in Ulster.

Both glacial sediments and the softer sediments on which they may lie are often disturbed by cryoturbation, the growth and melting of ice within the soil during permafrost periods.

Towards the end of the Quaternary,

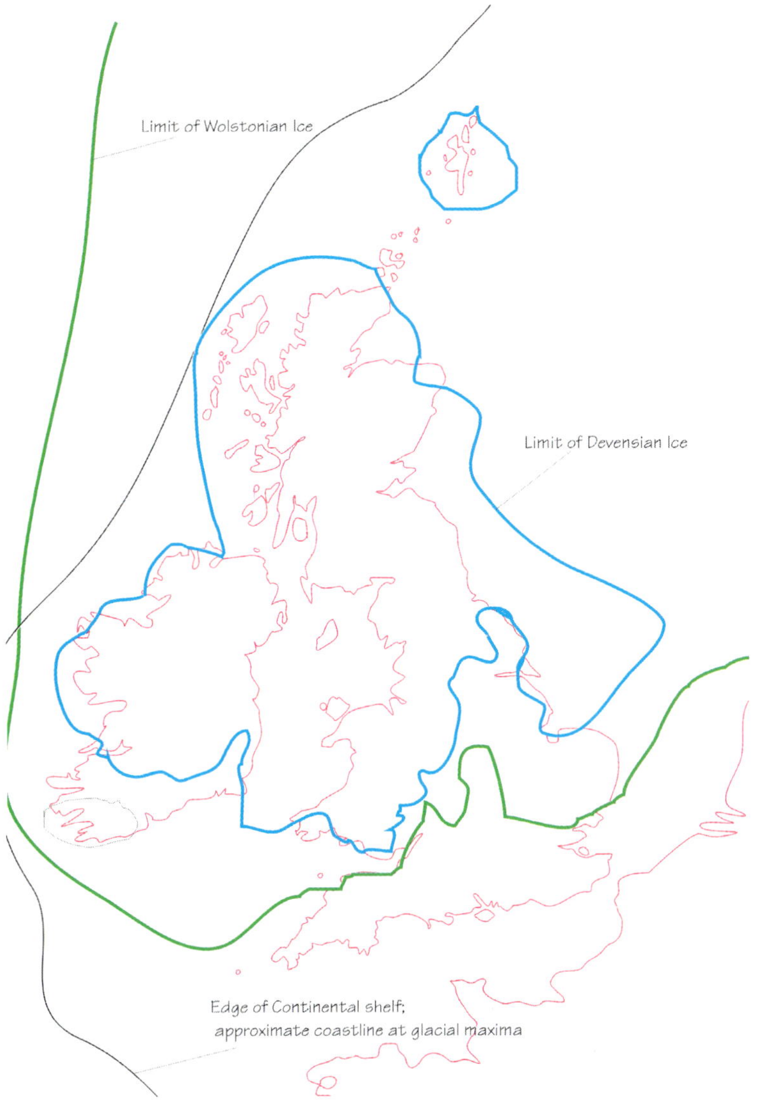

Fig.98 (top right) Limits of the last two major glaciations. Fig. 99 (below right) Cemented fragment of Pleistocene raised beach resting on Carboniferous Limestone, Gower.

Britain finally became separated from France, which resulted in a total rearrangement of marine tidal and other currents which continues to affect the distribution of continental shelf sedimentation.

North-western Europe, including the British Isles, is now a seismically quiet area, which can be predicted with some confidence to last for millions of years. Britain is presently thought to be in an interglacial, which started about 10,000 years ago.

While general sea level is rising due to melting of the former ice sheets, northern and western Britain, which had been most heavily loaded by ice masses, rebounded isostatically and this localised movement has more than compensated in many areas for the general sea level rise. There is some discussion of the significance of this in relation to marine flooding, etc, in Chapter 13.

Chapter 9 Landslides and mass movements

Introduction

Mass movement is the term that describes the movement of soil or rock downhill under the influence of gravity. It includes both rapid and slow phenomena. The latter include soil creep, whereby soil creeps inexorably but very slowly downhill as a result of physical processes such as freeze-thaw. Such movement is shown, for example, by the lean of fence posts downhill. However, the most spectacular and damaging manifestation of mass movement is landsliding.

Landslides

Landslides (Figs.100, 101, 102 and 103) are extreme cases of mass movement that occur when rock, soil or waste become unstable and move under the influence of gravity on a short-term timescale of weeks, days, or hours — that is, on a timescale that falls within a human frame of reference. The landslide usually has well-defined lateral boundaries and an underlying, or basal, shear surface or surfaces. As the material slides or flows, it may engulf buildings, bridges, roads and people. Landslides may be triggered by earthquakes, volcanic eruptions, heavy rain and flooding. Moisture between particles in soil or rock reduces its resistance to sliding, and hence prolonged or intense rainfall is the most common trigger. The sea or a river, nibbling into the foot of a slope, can remove support, as can the unconsidered actions of man. Earthquakes may induce failure by changing pore pressures within soils or rock. Volcanic activity can push the ground upwards and cause landslides.

Prolonged rainfall, acting on potentially unstable sections of the ground, has been shown in numerous studies to be the critical factor that sets off final collapse. Many recent landslides, however, are not new phenomena but are reactivations of much older (often ice-age) slides.

Major landslides within the UK are, fortunately, rare but nevertheless there are many landslides that have caused major damage to buildings and other property. Typically, these fail to make more than local headlines, if that, so that the true extent of landslipping is generally unappreciated. Some landscapes are more vulnerable than others: classically, where a permeable formation overlies an impermeable formation. Typical examples include the predominantly sandy formations of the Lower Greensand, which overlie Weald Clay; limestone formations of the Cotswolds, which overlie clays (Bredon Hill, an outlier of the Cotswolds, is completely encircled by landslides); and the Pennant sandstones of the South Wales coalfield, which overlie impermeable shales.

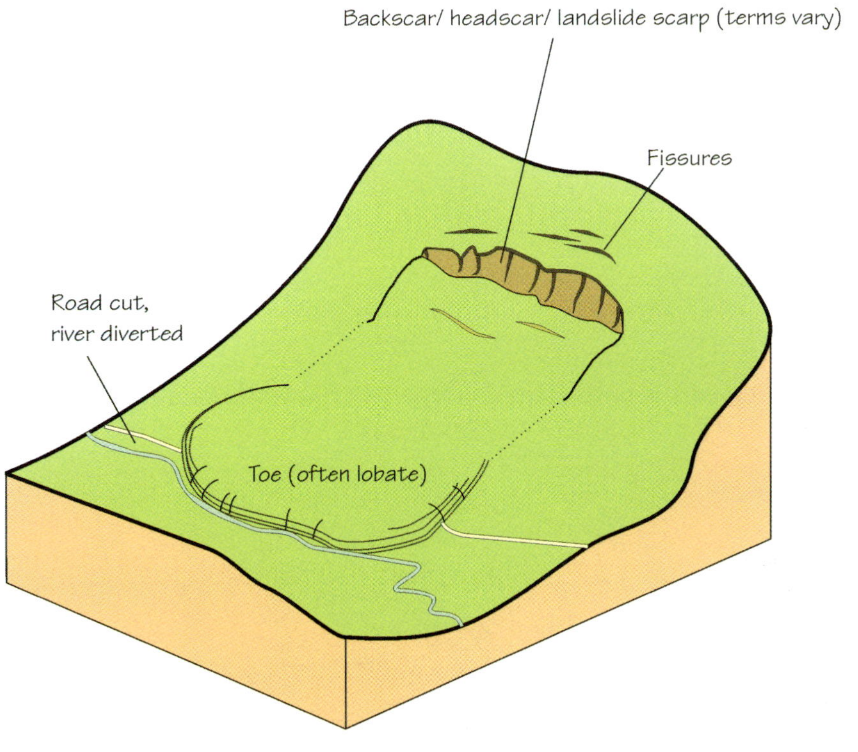

Fig.100 General features of a typical landslide.

Fig.101 The toe of this recent slip in London Clay, on the Isle of Sheppey, shows wrinkling that indicates a degree of fluidity at the time of the slip. The stony objects on the right, for scale, are nodules about 20 cm across. The road above this cliff is called Sea Approach; the sea is indeed approaching, but the irony was doubtless unintended.

All these formations tend to outcrop in hilly areas (but it is an old engineering geology joke that the only safe slope in clay is 0°). Geological Survey maps often show some landslips, but smaller or older ones may not have been mapped.

Other vulnerable scenarios include cliffs, banks, etc, that are under constant erosion from the sea or rivers and man-made features, such as road embankments or cuttings (Fig.104), which have not been properly designed.

Mining, particularly in hilly districts (Fig.105), can also destabilise slopes and contribute to, or directly cause, landslides.

It was realised during the 1960s that ill-considered development often created the circumstances under which landslides could more readily be triggered by other factors. A particular stimulus was the 1966 failure of Tip No 7 at Aberfan (see box on page 100), after which much more attention was paid to such issues. By that time, the improvements in knowledge and

Fig.102 Perpetually sliding cliffs of lower Jurassic clay, shale and mudstone, Dorset.

Fig.103 In this large reactivated landslip (Blaencwm, South Wales), the headscar is obvious, as is the hummocky and terraced ground forming the toe of the slip. The size is given by the immature conifers on top of the hill. Note the assisted drainage by pipes, left.
Fig.104 (below) Classic mode of slope failure following human interference: schematic sections.

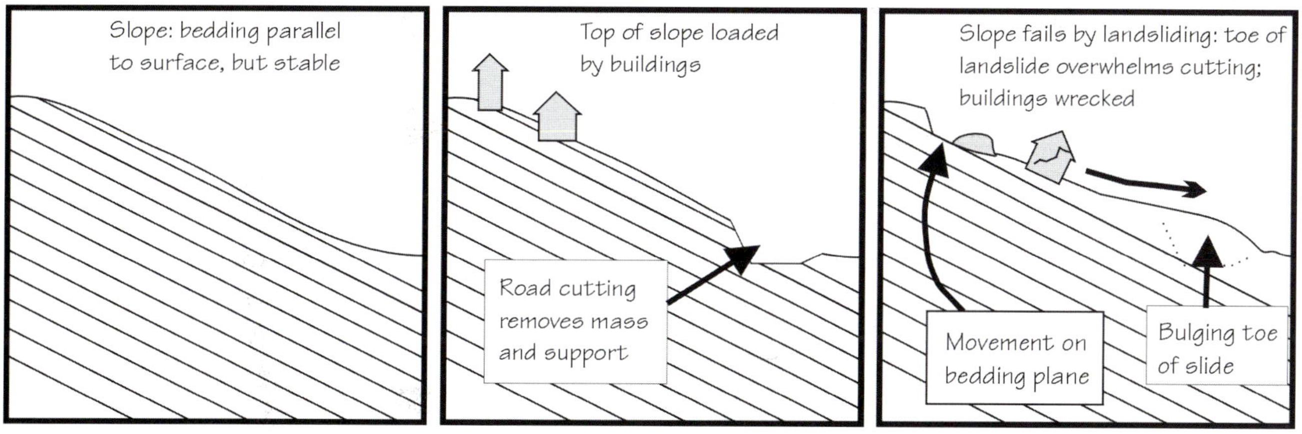

engineering techniques in the preceding 20 years or so made it possible to institute much more effective studies and remedial or preventive techniques.

Major British landslips include:

Folkestone Warren, a multiple rotational slip crossed by the Folkestone-Dover rail line. Here slices of slipped chalk overlie clay. The railway track has to be realigned on a regular basis.

Fig.105 (right) Surface cracking in hilltop above coal extraction area – this is often found above the head of landslides (compare Fig.100, p73).

Fig.106 Mam Tor: the abandoned road, with immense thickness of road metal; note hummocky topography beyond.

Mam Tor, Derbyshire (Fig.106). A slide 300m wide and 1 km long, thought to be over 3,000 years old. The local authority eventually gave up realigning the main road that crossed the slip and closed it in 1977.

Landslide categories

Geomorphologists and Engineering Geologists have sought to classify landslides as:

- Falls: collapse of steep slopes (usually rocky), where the initial movement is of a freely falling object.
- Slides: material moves along a distinct movement surface or slide plane.
- Flows: material with (usually) a high water content moves like a liquid.
- Topples: material becomes detached from a near-vertical face, the initial movement being outward rotation.

However, many landslides are very complex and may include all of these elements, although geotechnical investigation may be necessary to identify them. The general appearance of a landslide complex is shown schematically in Fig.100.

Recognising old landslides

Old landslides form particular hazards since they are easily reactivated. Most Ordnance Survey maps do not contain enough information to distinguish landslide forms. They can, however, often be guessed at from site features:

- Irregular hummocky ground on slipped material.
- Lobate or linguate form of debris and flows.
- Unsorted debris exposed in trenches or cuttings.
- Concave upper slope, convex lower slope.
- Head scars: more rounded in softer material, angular in hard rock.
- Anomalous dips in slipped masses, caused by rotation.
- Boggy hollows or linear ponds; wet ground at presumed toe.
- Mismatch of vegetation due to difference in soil porosity, etc.

It is, however, **essential that further research is carried out before** firm conclusions are reached.

Isle of Sheppey

The London Clay Cliffs of the Isle of Sheppey (Figs. 49, 101) face the North Sea obliquely and are therefore exposed to waves of significant power. This combination of soft rocks and quite intense marine attack means that the cliffs have been slipping constantly as far back as records extend. As new landslide toes are formed, the sea erodes them away, removing support so that further landslides can occur. These cliffs have been studied in detail since 1971. The mean rate of retreat since 1867 has been 1.6 metres/year. A number of residential areas are on top of the cliff. Many properties on the clifftop now have a very short life expectancy.

'Hidden' landslides

In many cases, local authorities or others have acted to 'cure' landslides by landscaping or by road repairs. This could mislead an unwary property purchaser or surveyor, who may not realise that the area is at risk (see Figs.107, 108 and 109). This is not to suggest that any authority is at fault, but to underline the point that a landslide-prone locality may not display all the symptoms.

In other cases, the presence of an existing landslide, which has reached a position of stability, is not recognised until it is cut into. Commonly this occurs when the toe is removed, perhaps for road construction. The change in loading can remobilise the landslide with catastrophic results.

Fig.107 Spot the old landslide: Franklands Village, Haywards Heath, West Sussex. Houses had to be demolished following a landslide into the stream valley.

Fig.108 (left) The Leith Hill landslip system reactivated late 2000. Fig.109 (right) The same location after repair a few months later by the highway authority: note distinctive tree pair at rear of view.

Special case study: Lyme Regis – a town on the move

Lyme Regis (famous for the film *The French Lieutenant's Woman* and for fossil ichthyosaurs) is one of the most (downwardly) mobile towns in the UK (Fig.110). The town's population is in the order of 14,000, and there is considerable tourist trade (ironically, much of it associated with fossil collecting), so that the potential for financial loss, as well as disruption, is considerable.

The town centre focuses on the steep valley of the little River Lym. The coastline, which rises steeply from the shore, is partly protected by the harbour and by a system of sea walls and groynes, forming all together an assemblage of inconsistently planned and ageing structures. To either side of the centre, the town is largely built on a series of ancient sea-cliffs and degraded (and therefore less obvious) landslides. Rising sea level and the absence of sea defences have permitted rapid cliff retreat. To the west of the town, the sea advances at about 1 metre per annum. There is a long history of storm damage on this exposed coast: in 1377 80 houses were destroyed.

continued.....

Special case study – Lyme Regis *(contd)*

Fig.110 Lyme is dominated by the adjacent high and unstable cliffs: view north-east over the town. Note yellow capping of Cretaceous strata.

Fig.111 Lister Gardens: path cracking and (rear, just visible) leaning lamp-post.

The geology of the underlying rocks comprises the lower Jurassic 'Lias'. This is formed mainly of shales and mudstones but with considerable quantities of tabular and nodular limestones, known as the 'Blue Lias', in the lowest beds. The cliffs are capped by unconformable Cretaceous Upper Greensand sandstones (Fig.110), which transmit water to the much less permeable Jurassic strata.

To the east side, where a complex system of landslide extends about a kilometre inland, the rapidly retreating Black Ven cliff long ago destroyed the former coast road to Charmouth, and the present Charmouth Road, which runs almost north-south, could conceivably be threatened before very long as the cliff retreat proceeds. Recession is removing the toe from an old, large slide that runs parallel to the Charmouth road. The landslides on this side of the town formed prior to the construction of the town's sea defences, but reactivation of old failures, and new, 'first-time' slips continue.

The dip of the underlying rocks (mostly lower Jurassic mudstones, limestones, sandstones and clays) is towards the sea – a classic risk situation that provides plenty of potential movement surfaces. The weak shales and mudstones are clearly, in most cases, the preferred location of shear planes.

The long history of movement, in some instances, has forced demolition of properties and wrecked earlier remedial measures. Movement is virtually an everyday occurrence, with cracks (Fig.111) being patched and car parks resurfaced (Fig.112), etc, on a regular basis. The local football pitch is bulging and one corner is protected by gabions (wire baskets filled with rock) (Fig.113).

In January 2003, a further small but significant movement at Church Cliff (Figs.114 and 115), involving loss of garden areas, gave rise to considerable concern and this has further devalued properties.

The local authority have maintained a close watch on the situation, with specialist input from various consultants and their in-house Engineering Geologist and Oceanographer. Detailed geological and Geomorphological mapping has allowed recognition of the individual components

Fig.112 (top) Charmouth road car park: landslide toe at north edge. Just above the gully grating, the very small black and white object is a monitoring point.

Fig.113 (above) Gabions overturned by pressure from landslide, entrance to allotments, East Cliff.

Fig.114 (right) January 2003 fall at Church Cliff: black Lower Lias shales exposed.

Fig.115 Close-up of Church Cliff damage: leaning trees and teetering fence. Dwellings are just beyond. Very little movement would be needed to cause catastrophe.

Lyme Regis *(contd)*

of the landslide systems. Study of historical development has permitted assessment of the effect of harbour construction, limestone extraction etc, (even in Victorian times the cliffs were being attacked, for cement manufacture, in this way despite their instability).

Inclinometers and monitoring points are installed at numerous locations and in the Lister Gardens park (Fig. 111), on the west side of the town, artificial drainage has been installed. Careful study has provided enough information for drainage to be targeted at the weakest horizons of the Lower Lias (the 'Blue Lias'), where water and shear planes are concentrated. Recognition of the linkages between the landslide systems has allowed the construction of 'event trees' with some predictive power.

In addition, residents are strongly encouraged to report new cracks or other evidence of renewed movement.

The West Dorset District Council even has a Landslide Warning web page.

Southend-on-Sea

Southend is built on low (about 30 m) cliffs which, historically, have been relatively stable. However, since 2000 (and almost certainly stimulated by the wet autumn of that year) the cliffs have suffered considerable movement. Lower Pleistocene permeable sands and silts overlie the much less permeable Eocene London Clay, producing a classic risk scenario.

Remedial work by the borough council cost about £0.5 m, including modified drainage, in order to preserve the large bandstand from potential destruction. However, in November 2002 a much larger slip occurred and closed part of the promenade dual carriageway. Property owners on the cliff top are naturally alarmed. At the time of writing, investigations to find the positions of the shear surfaces continue.

Rottingdean, Sussex, 1987

In October 1987, the sloping fields on the chalk downland to the east of Rottingdean, Sussex had just been sown. After days of heavy rain, culminating in a fall of about 60 mm in 12 hours, the saturated soil liquefied to a mudflow and swept rapidly downhill, enveloping a car on the nearby road in about a metre of mud. The flow then passed through parts of several residential streets, moving cars and overturning walls before running out of energy.

About 200 tonnes of soil per hectare was removed, and about £500,000-worth of damage was caused.

Mudflows can travel long distances. In this case it is noteworthy that only the (normally dry) valley bottom was affected by the flow. In hilly, and especially cultivated, areas the risk of such an occurrence should be borne in mind as probably small, but real.

Cambering, valley bulging and gulls

In relatively hilly areas and where the sedimentary strata are more or less horizontal, there may be a curious situation whereby relatively hard strata capping the interfluves seem always to dip into the valleys, no matter what the regional dip may be (Fig.116). This phenomenon is termed 'cambering'. It is thought to be due to the semi-plastic behaviour of softer underlying strata (typically, clays) which have been squeezed out from under the heavy, more rigid, overlying strata. Some authorities consider that this occurred only during the relatively recent ice ages, although there is no reason to suppose that such plastic flow should be confined to the ice age periods.

In some parts of the East Midlands, cambering has given rise to a depression of up to 30 metres below the 'expected' level of any particular stratum.

A related phenomenon is 'valley bulge', whereby valley bottoms in clay always seem to exhibit anticlinal dips. This latter phenomenon, however, is unlikely to affect buildings substantially as far as the writer is aware.

In the case of cambering, however,

Fig.116 Hypothetical cross-section (not to scale) to show cambering, gulls and valley bulging.

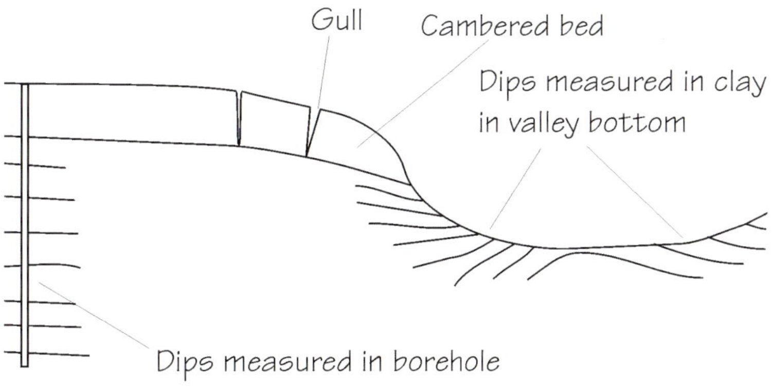

there is an associated development of cracks, more or less parallel to the line of the valley, which develop as the cambered strata start to tilt into the valley. These cracks, which are basically greatly enlarged joints, are called gulls (Fig.117). They may be of considerable depth, as deep as the thickness of a particular resistant bed, although they tend to fill either with soil or with the disintegrating remnants of any thin overlying strata. Such gulls can form the head of a landslip or a toppling failure in some circumstances, but as cambering appears to be a relatively slow process they are more likely to give rise to problems in connection with building foundations. There is, however, no firm boundary between gull development and incipient landslide.

In practice, few settlements have been built upon areas affected by gulls, so that the phenomenon is little known.

The surveyor should be cautious of the existence of gulls where property is built upon resistant formations that overlie softer mudrocks.

Fig.117 Multiple gulls in middle Jurassic Limestone (clay vale to left of view). In this case, the gulls are in massively bedded limestone and do not reach the surface: movement is absorbed in the more finely bedded and jointed rock above. In the field above the quarry face, there is no indication. Height of cliff face about 20 metres. Leckhampton Hill, Gloucestershire.

Chapter 10 Economic geology

The earth's resources

Introduction

Many of the earth's components have uses for building, can be smelted to make metals, or can be used as fuel, etc. They are extractive minerals. Such uses can affect the value of land:

As a valuable resource (e.g. gravel deposits on an otherwise impoverished farm).

As a nuisance (e.g. noise, pollution).

As a cause of instability (see Chapter11).

In the UK, the varied geology has given rise to most forms of extractive industry.

Advances in understanding of earth processes have made it possible to predict more accurately areas where certain minerals may be found. Britain may well have still-unguessed mineral resources, although whether they are economic to work will be another matter. It is certain that resources of tin, for instance, still remain in Cornwall but unless world tin prices rise substantially it will remain unworked. Gold is still thought to be present in parts of the Dolgellau gold area in North Wales, but again it may not repay the costs of exploration. On the other hand, political factors (such as the closure of overseas sources) could force the re-exploitation of even low-grade deposits in Britain.

Ore minerals

Ore minerals are economically valuable minerals that can be extracted, separated from valueless material, concentrated and processed to yield valuable metals. Other mention of these metals is given in the respective sections on Minerals and Mining.

Britain was recognised before Roman times as a source of tin, and has a long history of metalliferous mining, notably for lead, zinc, copper and tin. These resources were distributed unevenly throughout the Palaeozoic rocks of northern and western Britain. More rarely found were ores of antimony, nickel and cobalt. Silver was often extracted from lead ore. Manganese has been worked in the Abermawddach area of Wales, from Lamancha in Peeblesshire, near Tomintoul and other places, while a few Scottish localities bear small amounts of chromium, notably in Unst (Shetland) where it has been worked. Gold has been produced in small quantities from locations as widely separated as Avoca, Co Wicklow, Dolgellau in North Wales, and Sutherland in Scotland. Molybdenum was traced near Loch Shin by BGS mineral reconnaissance, and also in northern Scotland the Old Red Sandstone bears uranium which may be potentially workable. Wolfram, the ore of tungsten, has been worked near Dartmoor.

Coalfields often contain significant quantities of bedded nodular ironstone, which could be conveniently smelted by use of the coal.

Some Cretaceous iron ores have been extracted, while more recently Jurassic iron ores have been extracted (e.g. at Frodingham) by strip mining. Small quantities of ironstone were extracted from the Ordovician of North Wales in the past.

Coal and oil

These are discussed under Sedimentary Rocks (Chapter 4).

Industrial minerals

Britain's varied geology contains a number of non-metallic, non-fuel industrial resources, including:

- **Rock salt:** pre-eminent in the Cheshire Basin.
- **Potassium salts:** evaporite extraction in north Yorkshire, and used in the chemical industry.

- **China clay:** extracted from decayed granites in Cornwall; used for paper coating, paint fillers and in ceramics (Fig.118).
- **Ball clay:** for high-quality ceramics; extracted from the Eocene of the Dorset heathland, and at Bovey Tracey and Petrockstow in Devon.
- **Bentonite:** used for oilfield drilling fluids, cat litter and hydraulic barriers.
- **Hard rocks,** including granite, dolerite, hard limestones and hard sandstones, can be crushed to form concrete aggregate; some rocks are crushed further to make sand. There are widespread sources, mostly in upland Britain.
- **Dolomite limestone:** again widespread in parts of upland Britain; used in refractory manufacture.
- **Gravel:** gravel suitable for aggregate use is easily extracted from river valleys or offshore and needs minimal processing.
- **Brick clays:** widespread extraction both now and in the past for brick manufacture, both from clay formations such as the Kimmeridge Formation and from older Coal Measure mudstones.

Fig.118 China clay pit, Bodmin Moor.

Building stones; ecclesiastical geology

Building stones for roofing and walling

Many rocks have been used as building stone. A glance at the BGS map 'Building Stone Resources of the British Isles' shows a very large number of extraction sites even though only the larger and/or more important quarries are shown.

Trade usage of terms differs from geological usage. For instance, in construction 'marble' is used for any limestone that will take a polish. Consequently there is the Hopton Wood Marble, from Derbyshire, which is a black limestone; and the *Paludina* 'Marbles' (=Sussex Marble) of the Weald, which are shelly limestones (the eponymous snail fossil, originally called *Paludina*, has been renamed *Viviparus infracretacicus*, but the name of the rock remains unchanged to avoid confusion). In general, the trade regards most igneous and metamorphic rocks as one group, 'granite', while sedimentary rocks and 'marble' are two others. The properties of the stone, rather than its petrology, determine its building classification.

As an aside, the 'marble' fire surrounds often referred to in property particulars are frequently artificial products.

Until very recently building stone was a bulky product of low value that could not be economically transported very far. Consequently, in earlier times, except for a few exceptionally prestigious projects that only the nobility could afford, the nearest available building stone was used, and a myriad of tiny quarries existed that were often opened for the construction of a particular building. This tended to lead to limitations in construction style due to the properties of the rocks, thus giving the rich variety of British vernacular styles.

In older highland areas, large lumps of granite, dragged from the fields, or flat slate slabs gave a rough appearance to the local buildings. Castles or other large buildings were often built of the local material, even though it might be intractable to work, in order to avoid the vast cost of haulage. Antrim's Carrickfergus Castle is built mostly of the local Eocene dolerite and basalt.

Over much of south-eastern and eastern England the only durable local material was flint, which can be laid neatly (Fig.119) but, as surveyors will know, cannot be used very effectively at corners; hence the circular church towers of East Anglia. In Wiltshire and other Chalk areas, hard sarsen stones were broken up and used successfully.

Many older parts of settlements are characterised still by the local material: the Pennant Sandstone of South

Fig.119 Carefully knapped flint trimmed with oölitic limestone quoin stones.

Wales (used, for example, in Caerphilly Castle) and the Bristol area; the various oölites (Fig.119) of the 'Stone Belt' from Dorset to Yorkshire; or the brown sandstones of Glasgow (for instance, the Giffnock and Bishopbriggs sandstones of the lower Carboniferous) and the grey granite of Aberdeen. The often surprisingly bright red Triassic sandstones of the English Midlands are a visually satisfying part of the built environment.

Christopher Wren, however, was able to choose Portland Stone readily for St Paul's Cathedral at an early date, as the stone could almost be lowered straight into a boat.

In the south and east, relatively unsuitable stones (but still more durable than timber) were pressed into use, including Upper Greensand siltstones and the harder types of chalk. The latter are rarely left exposed to the weather, however, because of their generally poor durability.

Especially with the improvements of canal and rail transport, it became economic to select suitable stone from further afield. Thus Welsh slate soon roofed half of England. Other roofing slates, in the strict sense of a metamorphic rock, have come from Cornwall (notably Delabole), the Lake District, and Swithland in Charnwood Forest near Leicester. Exotic stones became readily available, such as the red granites of Shap Fell in the Lake District, or the imported but very attractive 'Norway Porphyry' (Larvikite) (Fig.120) which adorns so many bank façades.

Fig.120 Although some of the beauty of Larvikite can be seen in this picture, only the real thing can be fully appreciated.

The 'Stone Belt', which forms the Cotswolds, Cleveland Hills, etc, has long been the source of many building stones (including many oölites) which combined workability with reasonable durability. It has, consequently, been much exploited for the houses of gentlemen, churches, and public buildings. The Palace of Westminster includes much use of Clipsham Stone (from the Jurassic Lincolnshire limestone) as well as Anston Stone from the Magnesian Limestone of Yorkshire. The older buildings of Oxford University benefited from proximity to the Cotswold building stones: the Radcliffe Camera, for instance, is built of Taynton and Headington stones.

Local stones can be found not only in the walls of buildings, but also on roofs (not counting the ubiquitous slate). Fissile (splittable) rocks from many areas have been used, but several stones (mostly oölites) from the Cotswolds are quite well known, notably the Collyweston 'Slate' which can split to about 1 centimetre thickness. Purbeck 'slates', seen in Corfe, Dorset are quite heavy at about 1.25 tonnes per $10m^2$. Ham Hill 'slates' from Somerset are extremely durable and can last, reportedly, over 400 years. In Sussex, the Lower Cretaceous Horsham Stone, which is also used for roofing, is a sandstone, weighing about 1.5 tonnes per $10m^2$. In northern England many Coal Measures and Millstone Grit sandstones are fissile enough to be used for roofing.

While many building stones can be identified only loosely (except in certain cases), patient petrological work has often narrowed them down. This is especially the case with buildings that have considerable historical value, and allows deductions to be made about various stages in the construction of the building. Most such work has been done on churches.

Finally, although the structural use of stone has almost ended, natural stones still find significant use in

Fig.121 Contrasting use of stone: shelly limestone coping to wall of grey granite (note one very large felspar phenocryst).

cladding, especially for the more prestigious buildings. Massive masonry still has a limited use, as in seawall features and the like (Fig.121).

Ecclesiastical geology

Strictly, this discipline ought perhaps to be termed 'mural geology'. The field is important to architectural historians, conservation surveyors, and archaeologists, as well as being of interest to geologists. Collaborative work between the disciplines has enabled the identification of different phases of building repair and alteration, which may be correlated with surviving records and can indicate trading patterns and economic and political links. The ancient (eighth century AD/CE and later) walls of Brixworth Church (Northamptonshire) have been reported to show no fewer than 30 different types of stone, which can be traced to locations as far as 35km away.

Numerous buildings in southern England display French stones (notably Caen Stone and limestones from near Boulogne and Paris), and some of this has probably been recycled from former Roman construction.

An amateur geologist husband-and-wife team (this is a field where the amateur can be as valuable as the professional) worked assiduously to identify the use in time and space of Lavant Stone, a hard chalk quarried for a limited period near Chichester, and this information was useful to local and architectural historians.

Geology and concrete failures

One of the most common mass uses of hard rock is as aggregate in concrete. It has been known for some time that certain aggregates are unsuitable. Aggregate must be of low porosity and stronger than the concrete as a whole. However it may contain minerals with unacceptable qualities. Minerals implicated in various failures include pyrite, pyrrhotite (another sulphide), marcasite, gypsum, anhydrite, barite, aragonite, dolomite (especially Scottish dolomite) and various silicates including some forms of quartz, which can react with the alkaline cement to form a gel (alkali aggregate failure).

Sulphate in clay causes unprotected concrete to swell and crack. Reactions with clays can also give rise to growth of the little-known but deleterious mineral thaumasite (a complex hydrous calcium silicate carbonate sulphate).

The cement in concrete reacts slowly with carbon dioxide from the atmosphere to produce calcite, which can then be redeposited as a stalagmitic growth (Fig.122).

We have referred elsewhere to the problem of mundic blocks. Other aggregates including waste, such as colliery waste (1920s clinker breeze), can also be aggressive or disruptive; these can often be traced to particular production plants.

Fig.122 Concrete on a 1960s university campus showing stalagmitic growth.

Chapter 11 Worked and mined ground

Introduction

The RICS 'Red Book' has a surprisingly long section (Annex B to Guidance Note 1) called 'Old Mine Entries (i.e. Shafts and Adits)'. However, this title is misleading in its limitation. The section is mainly concerned with the mine entries only, but it does contain useful guidance, including the statistic that there are about 150,000 recorded old coal mine shafts and possibly as many unrecorded. PPG14 has a useful but limited summary of the risks. Coal mining search forms request details of recorded disused mine entries within 20 metres of the enquiry boundary (discussed in more detail below). This may not be close enough – the cone of subsidence into an old shaft could affect properties further away.

In the UK, worked and mined ground can be divided into:

- Mined areas: mostly for coal, but in some areas metal ores have been significant.
- Quarried areas: hard stone, gravel.
- Tunnels, etc, for other purposes.

Mines

Mining, except on the tiniest scale, completely alters the state of the ground. Huge masses of rock are fractured and subside, and the intensive mining of the last 200 years (mostly in coal areas) has left behind hundreds of kilometres of disused roadways, thousands of shafts and hundreds of square kilometres of broken and distorted strata. Full, or nearly full, records are available only for mines worked after the later 19th century.

DO NOT ATTEMPT TO ENTER ANY MINE WITHOUT EXPERT GUIDANCE. HAZARDS INCLUDE ROOF FALLS, DANGEROUS ABANDONED MACHINERY, TRIPPING, UNEXPECTED DROPS, LACK OF OXYGEN, TIPPED TOXIC MATERIALS, AND DEEP WATER.

Mines are of various types (Fig.123), irrespective of the material mined, and range from large commercial operations of the 20th and 21st centuries to tiny operations run by a few men. The illustration summarises the main types. Also to be found are 'trials' – small exploratory excavations that found no profitable material. These are common in some upland and/or rural

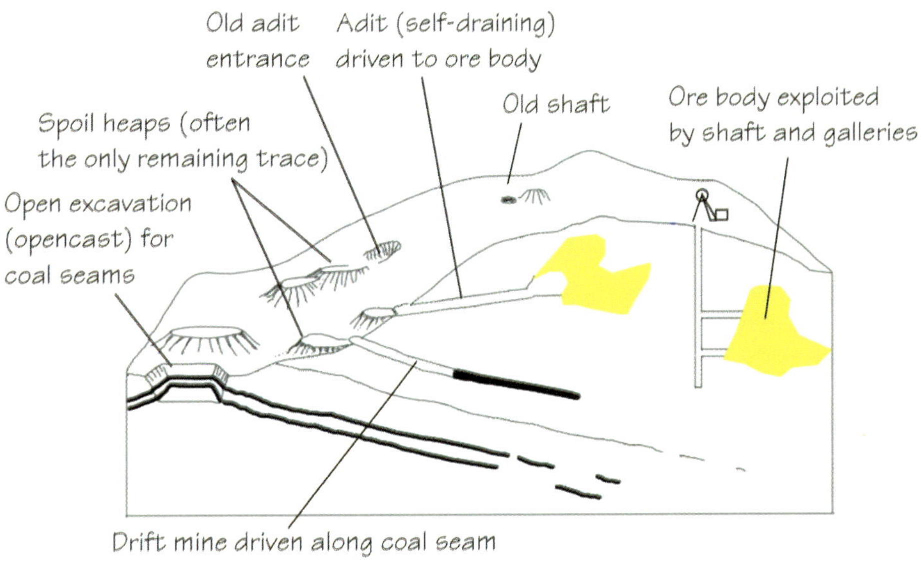

Fig.123 Mining features.

Fig.124 Old trial for metal ore, North Wales: one small spoil heap only. Figure, indicated, gives scale.

areas of Britain (Fig.124). Disused adits (an adit is designed to be self-draining) or drainage levels are often left open but barred. The water issuing from these tunnels may be toxic and discoloured. Streams issuing from shallow hollows may, in known or suspected mining areas, mark old adits that have partly collapsed. Toxic gases may emerge from some mines.

As well as subsidence, there is a risk of collapse into shafts or into crown holes (where the mine workings collapse internally, forming a void that progressively works its way to the surface).

When considering the environment of a property, study of large-scale maps (1:25,000 or better) often reveals useful clues that will start lines of research or, at the minimum, suggest a warning to the client. Obvious short, disused railway lines or trackbeds, for instance, especially in northern and western Britain, may have served an otherwise vanished mine, or a (possibly underground) military site, about which the Ordnance Survey has traditionally been coy (presumably in case the enemy of the day did not already know!).

Other clues on the map or in the field can include mining-related place names such as Drift Cottages or Incline Top. Older maps may be more helpful. Leats (long water channels leading from a stream or reservoir to the working site) may also suggest former mining activity, as may inclines. These are the former trackbeds of surface haulage ways for ore, coal or waste and are usually straight and at a constant angle.

Mining - coal

Coalfields occupy large areas of the UK (Fig.125), either at surface level or under a cover of later rocks. Deep mining has ceased in many formerly heavily worked areas, although coal is still being removed by opencast methods in some locations where it is economic.

Many coalfields are directly under large centres of population, for instance in central Scotland, Tyneside, Lancashire and Yorkshire, the West Midlands, South Wales and parts of Bristol. However, a number of smaller coalfields are more rural. The map shows a general indication of the layout of major British coalfields but there are many tiny outcrops that cannot be shown. Often, even the smallest have been worked to some extent and could still cause problems. In disused areas it is now rare to find a giveaway headgear (Fig.126).

Some of the older coalfields are already passing from public memory,

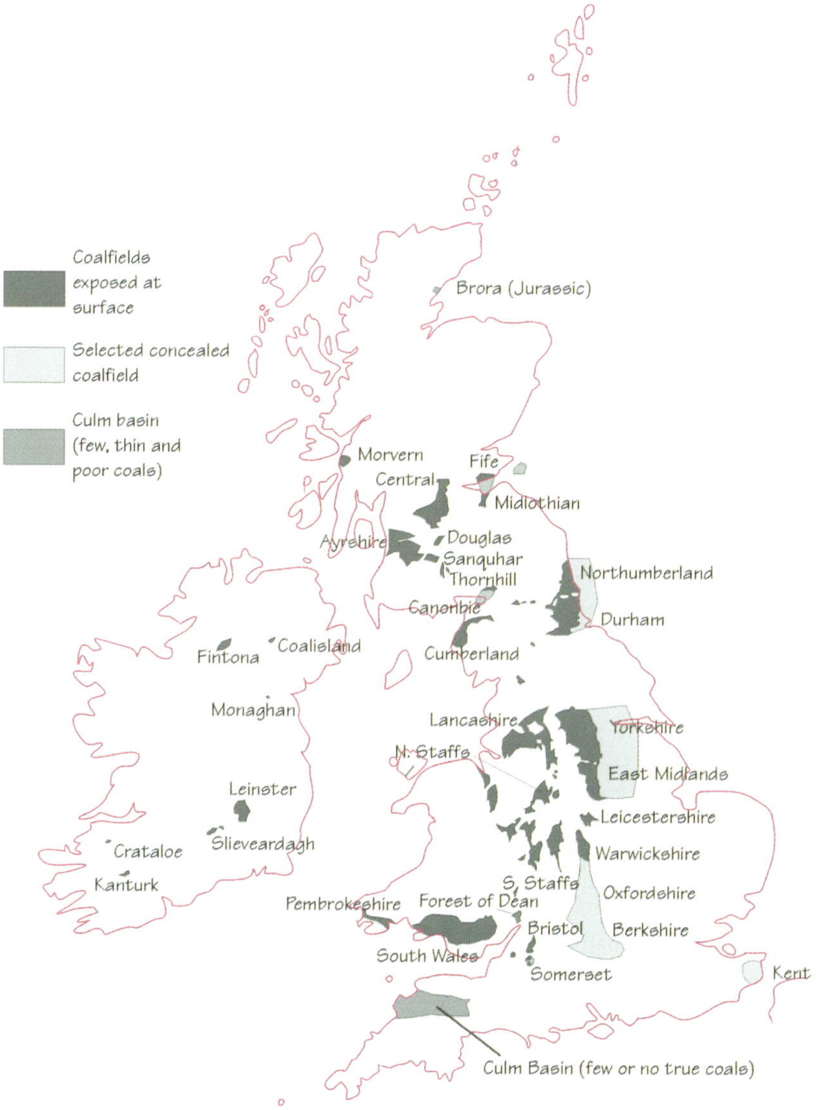

Fig.125 Generalised map of British coalfields. On this scale, only a broad indication can be given. However, the correspondence of coalfield areas with urban concentrations is immediately obvious.

Fig. 126 The preserved headgear of a 19th-century colliery.

and there are certain small coalfields whose records are now obscure. For instance, the Newent coalfield lies in a rural area north-west of Gloucester, and the Leebotwood coalfield is located in an equally rural area south-west of Shrewsbury.

In theory, Records Offices (in this case the Coal Authority) hold the necessary information. In practice, except for the more recent mines, records are often patchy or incomplete, and sometimes based on inaccurate plans. Some old shafts or other surface installations are named on maps (typically 'mine (dis)', 'old shaft', etc), and may be obvious on site (Fig. 127), but many are not. It is in these obscure locations that the greatest uncertainty may arise. Study of old topographic maps can be very informative in such cases. Place names are often a giveaway, as noted in the introduction. This is particularly true because of the association with urban or semi-urban areas. Bank, pit, incline, drift, coal (or cole) and the like may be found in various combinations.

Most British coal is upper Carboniferous in age (and, in Scotland and parts of Northern England, lower Carboniferous), but there are limited amounts of coal of other periods. In Devon, the deep Tertiary basin at Bovey Tracey, which is more noted for its high-quality clays, contains considerable quantities of lignite: in one borehole, a total of 150 metres of lignite was logged. Some lignite has been extracted in the past, and the basin remains the only major British reserve of lignite. At Brora, on the east coast of Sutherland (Highland Region), a one-metre seam of Jurassic coal was worked until 1974, while Tertiary coal has been mined from shallow workings to the west of Rochester (these have mostly collapsed).

Coaly streaks or thin seams are found in a number of formations, but are usually of no economic value and have not been mined, although prospectors have often sunk optimistic shafts (as at Bexhill, Sussex) which may occasionally be encountered. The extensive and much-contorted Carboniferous rocks of Devon (the 'Culm Measures') are not of deltaic origin and contain few or no true coals, merely a few carbonaceous seams (culm) that burn poorly and have been very little mined, in the Bideford area.

Coal forms layers, or seams, that are laterally persistent and are mined more or less horizontally (Fig. 128). In the richest coalfields, total extraction can amount to tens of metres of coal thickness. Although total subsidence is rare, partly due to the practice

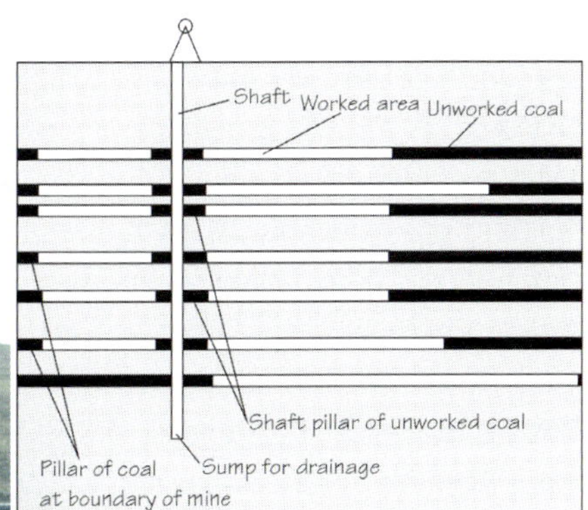

Fig. 127 (left) Some old mine shafts that have been capped in modern times are clearly labelled (British Coal notice on fencing in this case) and provide an immediate reminder of subsidence risk (and, in hilly areas like this, possible associated landslip risk). Nantymoel, South Wales.

Fig. 128 (above) typical method of coal extraction (schematic): in reality, seams are highly variable in thickness, are broken by faults, and may dip at high angles.

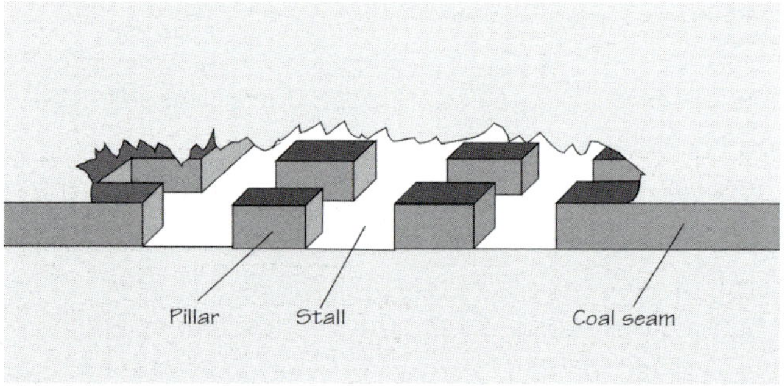

Fig.129 Pillar-and-stall working, with roof of seam notionally cut away. Schematic.

of storing waste rock (gob or goaf) in the voids left by extraction, the movement at the surface can be very substantial - over 15 metres is possible in areas of multiple seam working.

Old mines usually worked on the 'room and pillar' (or room and pillar, stoop and room) system (Fig.129), sufficient of the coal being left in position (pillars) to support the strata above. Over-extraction naturally led to a greater likelihood of collapse. In older coalfields with a long history of extraction, older properties often display distortion (Fig.130) caused by irregular subsidence. Many of these mines have collapsed already or been destroyed during opencast mining, but some remain. Old drift or adit entrances may also remain (Figs.131 and 132 show well-preserved examples).

More recent mines usually followed a process termed 'longwall' mining (Figs.133 and 134), with total extraction proceeding across a front up to about 400 metres wide. Extraction is performed in rectangular blocks termed panels, which can be 2 kilometres long. In such practice the face is protected by self-advancing steel support systems behind which the roof is allowed to collapse as the coalface advances. Deeper mining is usually less injurious in that the 'wave' of ground distortion caused by this process has a more indistinct surface effect. In practice a building passed under by such a face will first experience a heave-like process, with upward-widening cracks, as it 'leans' into the laterally expanding 'bowl of subsidence'; and then a reverse movement as it reaches the floor of the bowl. If there is a resistant sandstone or other bed at or near the surface, movement will tend to be concentrated at joints, giving rise to areas

Fig.130 (above right) This row of former mining cottages is in an area where deep mining has now ceased and further significant settlement is less likely. Notice that the middle cottages lean back, while the end cottage leans forward and sideways. Residents have straightened up some features, such as windows and porch tiling.
Fig.131 (right) Abandoned drift mine entry hidden in woods (but long known to the author). The nearest house is very close and is just visible among the trees. South Wales.

Fig.132 (above)
Some drift mine entrances are semi-permanent.

Fig.133 (above right)
Longwall mining: above and behind the operative are the self-advancing roof supports. As these advance, the roof collapses behind them.
The coal-cutting machine on the right advances to the right in successive cuts, taking its conveyor belt with it. After about a metre has been cut from the face, the props are moved forwards.

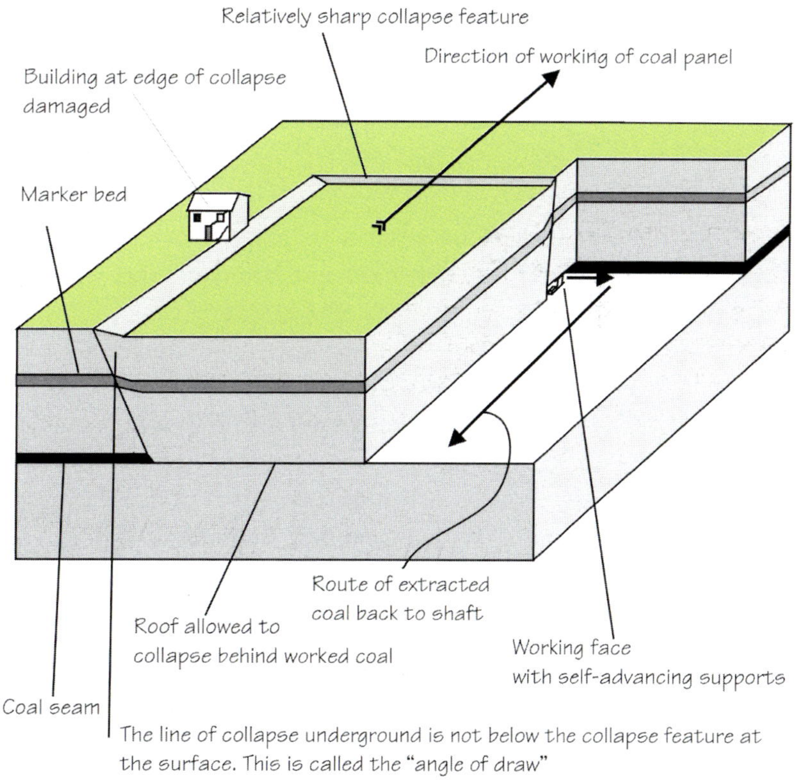

Fig.134 (right) Longwall working (the ground is notionally cut away on the right). Schematic. Compare with Fig. 133.

of concentrated stress and even fissures. Where the seam is thick and the panel is shallow, the worst damage will be caused.

In the longwall system, the ground usually stabilizes within the year. Any damage caused by the subsidence should be repaired under the provisions of the Coal Mining Subsidence Act 1991. The cost is incorporated in the working budget. However, recent buildings constructed without recommended precautions may not be subject to compensation.

The danger to buildings lies in uneven subsidence that distorts the building and/or its services, rather than in subsidence *per se*, which merely lowers the ground.

Walsall, Nr Birmingham

In the 1930s the Harrowby Road estate was constructed on a mining area. By 2002, more than half the properties lay empty following subsidence which made the houses unserviceable. In nearby Hughes Road, a few years earlier, a house disappeared into a mineshaft.

Not very far away, the M6 has subsided by about a metre between junctions 3 and 4 following mining from the Daw Hill Colliery.

Whitehaven, Cumbria

The Crowgarth Mines were reputedly so close to the surface that miners could hear the Town Hall clock striking. Whether this is true or not, subsidence caused great problems. In this particular location, proximity to the high sea cliffs meant that adits were driven from the cliff face and spoil was tipped over the top to be washed away by the sea. All this led to increased instability of the cliffs, which has required remedial action.

The Barlaston (Staffordshire) area was shown, in a recent engineering geology paper, to be subject to fault reactivation as a result of mining and this phenomenon will result in distortion of a theoretical subsidence pattern.

In some areas, seams close to the surface were worked, in the earliest times, from bell pits. Shallow shafts were sunk to the seam, which was then worked in all directions until serious instability set in. These have usually long since caved in, and leave depressions that may follow a pattern that indicates the outcrop of the seam.

As well as subsidence due to extraction, problems are frequently caused when old shafts reappear. Modern thinking is that 30 metres cover above old workings is generally the minimum safe margin. However, the capping or filling of older shafts is often of poor quality (for instance, supported on a staging constructed of wood). Capping style varies regionally – for instance, in the West Midlands 'beehive' caps were often used, where a brick dome was constructed over the shaft head below ground level, and then hidden by fill. Shaft fills can fail in several ways (Fig.135), leading to sudden catastrophic opening-up in unexpected locations. A large zone around the shaft head can be affected.

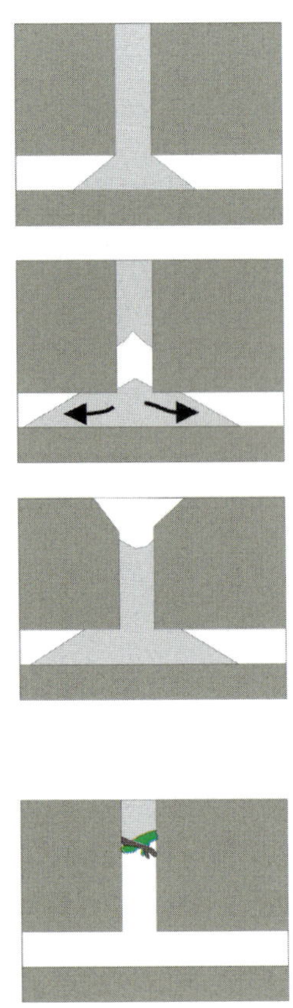

Fig.135
Possible types of mine shaft fill: schematic sections.

Upper three pictures: shaft filled by tipping. Base of tipped material later settles by gravity, allowing fill collapse and caving of shaft edges.

Lower picture: fill tipped on top of perishable barrier such as tree trunks, corroding metal

Coal mines where coal extraction has ceased may still have uses. At the now closed Prince of Wales Colliery in Pontefract, methane seeping from the extensive closed workings will be burnt on site to create electricity (see below, Coal Mine and Coal Bed Methane). Such uses may camouflage the existence of the former mine.

Mine drainage tunnels (drainage adits or soughs) were often driven to allow gravity drainage of older shallower mines in hilly districts (this also applies to other types of mine). In some larger mine systems, they were used for water transport (for instance, the Duke of Bridgewater's mines near Manchester). The tunnel portals often remain as low arches, usually with some sort of grating, in locations far from the original mine. Frequently they still function, with unpleasant-looking discharges.

Coal mines often have an orange or reddish discharge, stained by iron compounds (iron oxyhydroxides) derived from sulphide minerals, such as pyrite, that are associated with the coal in small quantities. When, as the result of the mining, the sulphide minerals are exposed to the atmosphere, they oxidize, creating acidic solutions. Such solutions are able to dissolve metals, such as cadmium, which are present in small quantities in the coal seams and adjacent sediments, resulting in a discharge with toxic properties. The reddish discharge should alert the property surveyor to the potential risks.

Coal mine and coal bed methane

Methane arises from decaying organic matter and is adsorbed into the fabric of coal. Mining (or faulting, etc) releases this methane, which continues to be emitted after the mine has closed. There are also enormous reserves of methane in unworked and economically unworkable coal reserves (coal bed methane).
The methane has been harnessed in the past on a small scale (an obvious use is to fire colliery boilers, for instance) but several projects have been devised to harness methane commercially.

At the time of writing, coal measure methane is harvested at the following abandoned mines:

- Silverdale, Staffordshire
- Shirebrook and Markham, Derbyshire
- Steetley and Bentinck, Nottinghamshire
- Hickleton, South Yorkshire
- Wheldale, West Yorkshire

And at the following working mines:

- Tower Colliery, South Wales
- Harworth Colliery, Nottinghamshire.

Numerous licences have been issued for coal bed methane.

There are a few recorded cases where methane has leaked out of mines and entered property. Such problems are dealt with by the Coal Authority.

Mining - other

Underground extraction of metals, building stone, chalk and flint has occurred in many areas.

Metals

Metalliferous mining has occurred predominantly in the older, harder rocks of the west and north, mostly for copper, lead (Fig.136), zinc, silver, gold, and iron. Tungsten has been worked in Cornwall. Ireland, however, has only a limited history of metalliferous mining.

Workings are widely distributed, but are generally on a rather small scale. Because of the mode of occurrence of the minerals, the mines are different in character from coal mines (Figs.123 and 137). Most of the mines have been of limited extent and within rural areas, but in

Fig.136 (below) This Derbyshire lead mine was worked as late as the 1950s (by the tiny headgear in the centre).
Fig.137 (right) Typical mode of operation of metal mine (schematic).

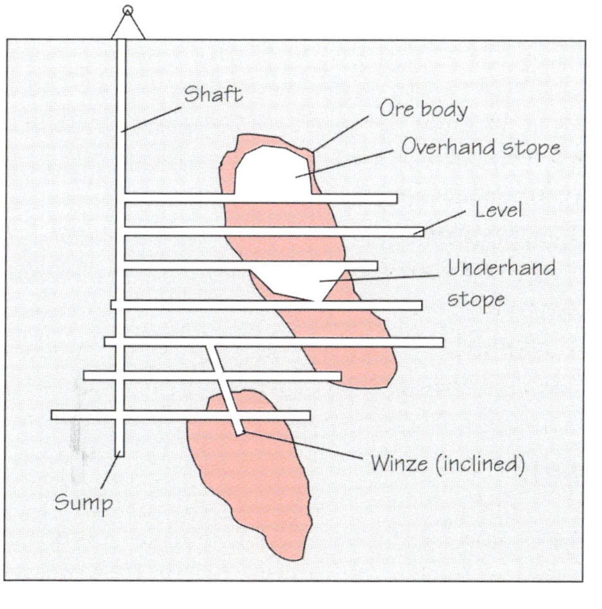

Fig.138 (top) Fortunately, unguarded mineshafts like this, in a remote location, are now very rare.
Fig.139 (above) This old metal mine level has been secured.

some cases substantial settlements have been undermined. In the large Derbyshire village of Eyam, for instance, an old shaft, still used for ventilation extraction, is located in the middle of the village.

Many mines, referred to in old records, have now been lost and remain as potential hazards. At Cusgarne School, in Redruth, Cornwall, tin workings were discovered under the playground in October 2001, and had to be grouted. Devon and Cornwall in particular are riddled with old workings, often near or under settlements. The Great Lode of Dolcoath Mine was worked out to a width of about 15 metres in places and to 1,000 metres depth. Although this is exceptional, it gives an idea of the size of the voids that may exist at depth.

Parts of northern England also have settlements that are thus affected, but to a lesser extent. Fortunately, many mines are still in rural areas and do not currently form a building (as opposed to a safety) hazard. Examples of abandoned mines are shown in Figs.138 and 139.

Metal ores are typically found in relatively narrow veins, although occasionally more compact deposits are found. Extraction leaves voids that correspond roughly to the shape of the ore body (usually an irregular sheet) and ore removal was often total, leaving perhaps a few wooden props to wedge apart great masses of potentially unstable rock. As the props rot, or other factors change, collapse can ensue. Because of the advantage of gravity in handling the heaviest material (the raw ore) many mines were worked upwards from the access galleries, often to within a short distance of the surface. This process is termed overhand stoping. The void thus left is a stope. The extracted material would be trammed or otherwise removed from the gallery at the base of the stope. Waste rock, or deads, would often be left on site, sometimes on stagings rigged across the timber supports. When these decay, the resultant collapse can be very dangerous.

Stopes (Fig.137) were often worked to within a short distance of the surface, and sudden and unexpected collapse can occur in such cases. Underhand stoping refers to working downwards, which had the disadvantage of clumsier handling unless a connection could be made to a gallery below.

In the case of the once-famous Snailbeach lead mine, in Shropshire, it is reported that tractors repeatedly broke through the thin cover over a stope that had almost reached daylight, and the site had to be made safe. Opencast mining at the vein outcrop has also been practiced.

In the Weald of south-east England, seams of nodular iron ore (clay ironstone, or 'mine') were worked until the 17th century from bell pits, up to about 12 metres deep. These have usually collapsed or been backfilled. Like coal bell pits, they often remain as damp hollows. Before building takes place on such sites, investigation and consolidation is necessary.

In the North Yorkshire Moors and parts of Northamptonshire, ironstone has been extracted from the Jurassic rocks by mining.

Industrial archaeology books are often the most accessible source for basic information on old mining areas, and there are societies that explore and map old mines. Some of these maintain web pages.

Salt

Salt is now mostly extracted by brining – pumping water into the salt beds and removing it after it has dissolved salt from the beds of rock salt below. Subsidence gives rise to surface depressions, or 'flashes' that usually fill

Fig.140 A modern salt mine.

with water to form ponds. Parts of Cheshire are extensively affected.

In Cheshire, in general, thick glacial deposits overlie the Triassic Mercia Mudstone Formation, which contains two halite horizons, each of 20-30 metres thickness and 100 metres deep or more. These deposits were at first mined (from the 17th century into the 20th century). Greed led to the provision of very limited support for the rocks above, and the old mines often collapsed spectacularly and flooded, with considerable damage and loss of miners' lives. Rather than pumping the mines out, the relative ease of mining led to the sinking of new shafts nearby. Some mine voids are thought to remain open and these are now brine-filled. As voids 10 metres or more in height were commonly excavated, the potential for damage is clearly enormous. On occasion, water-filled flashes have suddenly emptied themselves dramatically into collapsing workings below, leaving voids tens of metres deep.

Compensation may be payable under the terms of the Cheshire Brine Pumping (Compensation for Subsidence) Act 1952 as amended.

Areas affected include: Alsagar, Bucklow, Crewe, Congleton, Knutsford, Lymm, Macclesfield, Nantwich, Northwich, Runcorn, Sandbach and Winsford. Many of the mines may contain toxic material following decades of uncontrolled dumping, which adds to the hazard. Northwich is particularly at risk, with around 250 old shafts and 90 old mines in the area, like a giant ant's nest. The intention is now to fill the voids under Northwich with a grout of Pulverised Fuel Ash (PFA), cement and salt. Approximately 0.75 million cubic metres of brine will be removed and the salt content reclaimed, with a programmed completion date of 2007.

Cheshire is the only area now producing mined rock salt (most of which is spread on roads in winter), from a single large modern mine at Winsford operated by Salt Union (Fig.140). Large pillars of salt are left to maintain support to the roof; many years of reserves remain.

Gypsum

Deep beneath the wooded hills of East Sussex, near Robertsbridge, British Gypsum deep-mine gypsum from the Jurassic/Cretaceous boundary for plaster and plasterboard products. The mining is very discreet, within a wooded rural area, and casual inspection of the locality may not detect it.

Special case study: subsidence at Gilmerton, Edinburgh

Gilmerton, now a suburb of Edinburgh (Fig.141), lies on the western side of the Midlothian Coalfield and is underlain, below a relatively thin layer of glacial till, by the Lower Carboniferous Lower Limestone Group. This contains not only limestones, but shales, sandstones and a few coal seams. The rocks dip generally to the south-east (Fig.142). One of the limestone formations within the group, the Gilmerton Limestone, had been quarried in the area. Three quarries are shown on old maps.

Fig.141 Gilmerton: location.

During 2000, cracking and minor movements began to affect houses in Ferniehill Terrace, Gilmerton. On 16 November, a major subsidence event affected these houses and a second, larger, event occurred two days later. Four further events occurred during December. The pattern of the subsidence indicated a progressive failure.

Following the first event, the well-known engineering group, Arup Scotland, were called in by City of Edinburgh Council. Investigations began immediately, and intrusive borehole investigation had already begun by the time the third subsidence event occurred. Both Arup and the City of Edinburgh Council also undertook background and documentary research.

The following was found and forms a classic example of such an investigation, the findings of which are now in the public domain.

The Gilmerton Limestone had been extensively quarried during the 18th century. The lowest part of the limestone is a blue, high-quality limestone about 3 metres thick, which had been extracted preferentially. So valuable was this blue limestone that it had been extracted by mining galleries driven down-dip from the base of the quarries. These limestone workings were still accessible in 1940 and were indeed considered for possible air raid shelter use.

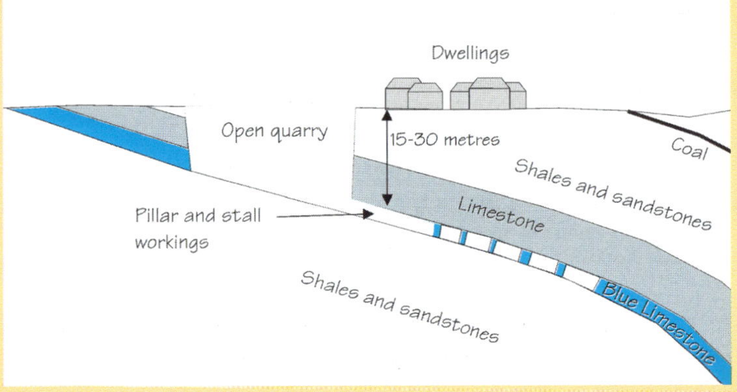

Fig.142 Schematic cross-section through the affected area of Gilmerton.

The limestone was extracted by the room and pillar method, leaving pillars "about 4 feet in diameter" [Arup repeating 1940 BGS report]. The extraction rate, however, was estimated to be 80-90% because of the strength of the limestone. In other words, only 10-20% of the limestone had been left to support the roof.

Workings ceased at about the end of the 18th century, as the water table had been reached. Mining began again in the 1820s, when the workings were drained by the opening of a colliery to the south of this area. The Arup investigation, however, showed that the workings are now flooded, presumably as a direct result of closure of the colliery and cessation of pumping. It was also found that the cover of rock over the workings ranged from 15 to 30 metres approximately (shallower closest to the quarry access).

After the Second World War, the open quarries were progressively backfilled. They had been totally filled by 1970 and the area redeveloped. However, it seems that no construction actually took place over the quarry sites. Site investigation in 1961 suggested the presence of cavities, as did further investigation in 1965. However, there is no evidence prior to 2000 of movement of the ground above the disused workings. Arup stated in their report conclusion: "The precise cause of the ground collapse after such a long period of relative stability will never be known for certain, but the evidence suggests that it was caused by the shear or compression failure of a single or

continued opposite page

limited number of stoops [pillars] in the area of the initial collapse. The collapse of these stoops resulted in additional load being transferred to the surrounding stoops which in turn collapsed. This process continued until sufficient resistance was met to halt the collapse. This resistance is believed to be the wall of rock following the limit of the open workings."

Arup further concluded that the rise in ground water level could have contributed to the weakening of the pillars as a result of preferential dissolution of the pure limestone in the pillars. They also noted that the mines had not closed up as a result of the collapse, and that there was potential for further ground movement within the affected area. In short, the area has now been sterilised for constructional purposes.

This information is abstracted from a copy of the detailed report, which was kindly made available by Arup Scotland.

It is known that other parts of the Edinburgh suburbs are likely to be affected by similar or related problems, and further investigations were put in hand following the reported matters.

Chalk

Chalk sounds like an unlikely material for mining, as it forms the surface in large areas of England. However, it has been extracted for use as lime from beneath a cover of later rocks, for instance beneath Reading, where in 2000 the sudden appearance of voids in Field Road made national news over a long period until the voids were grouted.

There are similar occurrences at Bury St Edmunds, Plumstead (South-East London), Frindsbury and other locations. The well-known Chislehurst 'Caves' are actually chalk mines. Small, isolated mines or 'dene holes' were also dug by farmers for lime.

Prehistoric flint mines are similar to dene holes (a dene hole is a vertical shaft with a chamber at the bottom), and occasionally new ones are discovered.

Building stone

Particularly in the Bath area, good-quality building stone has been extracted from underground galleries under the gently sloping plateau of the Cotswolds. Large areas in the Bathford, Box, Freshford and Corsham districts were literally undermined in this way and during the Second World War were used for ammunition storage (complete with rail access) and aircraft factories. They are now used for mushroom growing and secure storage in part, although some stone is still extracted, and most of the voids remain empty. At Combe Down, Bath, serious hazard has been recognised (see special case study overleaf).

There is a site of international geological importance at Wren's Nest, Dudley. A hill of Silurian limestones, with many excellently preserved fossils including the trilobite *Calymene blumenbachi*, the so-called 'Dudley locust', was quarried for centuries to provide building stone and agricultural fertiliser. Huge quantities were excavated during the industrial revolution to serve as blast furnace flux in the local iron and steel industry, leaving the hill undermined by a network of underground workings, which extend some 100 metres below the hill.
A number of other areas in the vicinity have been affected by limestone mining, including: Dudley, Sandwell, Wolverhampton, Walsall, and Telford and Wrekin.

Bixhead Stone Mine is part of the Bixhead Quarries near Coleford (Forest of Dean), where Upper Carboniferous Pennant sandstone was worked for more than 500 years, ending in the 1960s. Some of the larger chambers could easily contain a house.

Mine waters

Most deep mines extend below the local water table. During their working life, operation depends upon pumping but once the mine closes, water rises within the disused voids to the level of the water table or of any still-functional drainage tunnel (adit or sough). If these adits should become blocked, the hydrology of the area can

Special case study: Combe Down stone mines

The city of Bath, with its historic terraces of Georgian Bath Stone buildings, is situated in the valley of the Bristol Avon, which is flanked on both sides by steep hills. The hills owe their existence, in large part, to the presence of middle Jurassic limestone formations, which are resistant to erosion. The Great Oölite Formation forms the flattish hilltops, and the Inferior [in this sense, lower] Oölite is present at a lower level. Many of these limestones form satisfactory building stones.

Combe Down lies to the south of Bath (Fig.143). It is approached by steep slopes and has a flattish top. Here, the presence of the Great Oölite above the less resistant Fuller's Earth Formation of clays has led to cambering (see section on Cambering and Gulls) of the Great Oölite at the edge of the plateau, in particular, with numerous gulls. The surrounding slopes have suffered from land-slipping.

With the expansion of Bath in the 18th century, the extraction of the limestone became a major industry, and the quarries were extended by tunnelling into the hill. The workings were mainly active during the 18th and 19th centuries, although some workings continued until the early 1900s, by which time the useable stone was mostly worked out. It is thought that approximately 80% of the mines have less than 6 metres cover, while some sections have as little as 2 metres cover. The situation has been made worse by unsatisfactory mining practices and the 'robbing' of stone from the supporting pillars. No records of the workings were made prior to 1872 (when the Mining Act came into force).

Much of the plateau is now occupied by residential development. There has been a history of past collapses, including around some access shafts, but hitherto no major damage has been caused.

Since the mine closed, factors such as rainwater percolation and traffic vibration have exacerbated existing weaknesses, leading to fears of potentially widespread collapse. Following an underground survey in 1994, which was carried out by the then Bath City Council, the extent of the problem was realised and in 1999 the then DETR initiated a land stabilisation programme pursuant to the Derelict Land Act 1982. Parsons Brinckerhoff were appointed consultants and began to carry out the necessary site investigations, which included non-intrusive geophysical investigations, such as ground-penetrating radar. This is a satisfactory technique for some shallow voids.

A number of stabilisation options were considered and Parsons Brinckerhoff recommended the use of foamed concrete with limestone dust as a preferred infill material for the voids. However, the hill is significant in terms of water supply and it is intended that some area of the mines be filled with limestone chippings so that there is minimal interference with hydrology. Those areas which are significant in respect of bats and mining archaeology are to be supported by other structural methods.

At the time of writing, the scheme is subject to planning procedure and consultation. However, it was deemed necessary to carry out some emergency stabilisation. As part of this work, an underground roadway was constructed and this led to the discovery of a second, lower, mining level, the full extent of which has yet to be assessed.

As a result of all this, there are traffic weight restrictions, bus re-routings, and other problems, all of which affect the residents of the area. Remarkably, such is the popularity and attraction of Combe Down that there has been very little effect on property values according to local agents.

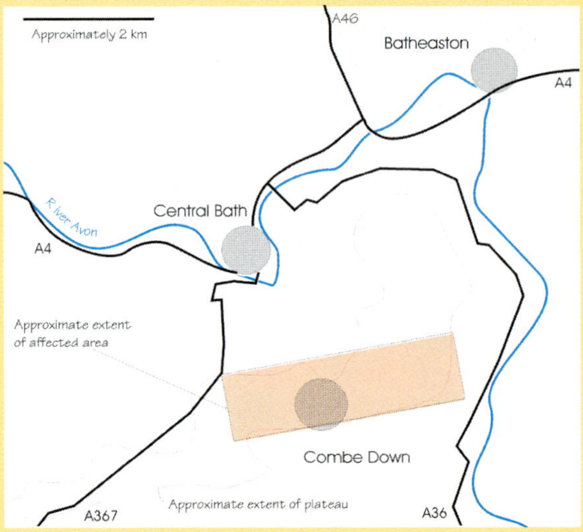

Fig.143 Combe Down: location.

be radically altered. This could give rise to seepage elsewhere that could undermine buildings or cause landslips.

In complex mined areas where the mine workings interconnect over long distances, mine water from uphill mines can issue from the shaft or other non-drainage entries of downhill mines (and also pollute adjacent aquifers). Modelling suggests, for instance, that in the Nottinghamshire coalfield, this may take place when abandonment finally occurs and pumps are switched off.

Spoil heaps

As waste material costs money to move, spoil has traditionally been dumped close to the mine access (Figs.144 and 145). In some cases, this has resulted in many millions of tonnes of potentially unstable material. The spoil from major (mostly coal) mines has, by now, often been levelled or regraded (Fig.146) and then topsoiled so that it now presents a wholly different appearance. Even so, once you have 'got your eye in', many such old tips can be picked out (Fig.147).

In the case of lesser mines, spoil heaps, although often overgrown, may still retain their distinctive forms.

Fig.144 (top left) This old coal tip shows gullying. Serious failure appears quite possible. There are houses nearby. However, the view appears unchanged from fig.145 (top right), despite the steep tip slopes.

Fig.145 (top right) 1974 view.

Fig.146 (left) Mine waste reclamation in progress.

Fig.147 (above) The large rather pale green fields in the middle ground represent reclaimed spoil/opencast areas. Hirwaun, South Wales.

The more toxic ones may still support little or no vegetation. Apart from problems of instability directly affecting the waste material (one need mention only Aberfan, the location of the 1966 disaster - see box), the tips are a sure indicator of potential underground voids in the vicinity.

Surface water runoff that has passed through or over the waste may pick up heavy metals and may be toxic, introducing another adverse element. In Devon and Cornwall, mine waste has been used in the past as a constituent of concrete building blocks. Some of this contains pyrite (q.v.), which oxidizes, expands and bursts the blocks apart.

Aberfan

Disasters such as Aberfan are always worth retelling, since the lessons are not always learned. Aberfan is on the west side of the Taf (Taff) valley, south of Merthyr Tudful (Tydfil). For years, colliery waste had been transported by tramway from Merthyr Vale Colliery shafts on the east side of the Taf. The waste was tipped high on the mountainside in a series of subconical tips. One of these, Tip No. 7, was partly on sandstone of the Brithdir Formation and partly on the mudstones below, where a spring emerged. There had already been a number of warning signs, including small failures and slides, before the tip slope failed catastrophically from excessive top-loading on 21 October 1966. The mass of sliding waste, moving faster than a running eyewitness, crossed a railway line. About 38,000 cubic metres of waste entered the settlement, destroyed a school and killed 144 people.

The entire incident was preventable.

Mining searches

In the relevant areas, purchasers should have special searches carried out by their legal advisers. This can be done for Coal, Tin, Limestone, China Clay and Cheshire Brine

These searches can reveal the answers to many of the issues discussed here, but you should bear in mind that the information may be limited and not consistent between the different types of extraction. The supplier of information will be forced to make various assumptions, and in many cases very old mines may be unrecorded.

For the most important mining records, coal, the Coal Authority warns, *inter alia*, that (summarised as at time of writing - do not rely on this information, and always check the up-to-date position):

- Not all information is in its hands, or information may be subject to updating.
- Older mine records may be inaccurate and may show confusion over seam names.
- Other minerals may have been worked in conjunction with the coal.
- Only those mine entries (estimated centre point) within 20 metres of the boundary of the property will be notified.
- Actual or conjectured faults at the surface which are not known to affect the stability of the property are not reported.
- Care is needed to advise the precise boundaries of the property to the authority.

The Coal Authority will report (summarised as at time of writing - do not rely on this information, and always check the up-to-date position):

- Past coal mining for which the property is within the zone of likely physical influence.
- If appropriate, whether ground movement should have ceased.
- Possibility of unrecorded workings.
- Present coal mining for which the property is within the zone of likely physical influence.
- Potential future coal mining.
- Shafts and adits within 20 metres of the boundary, with plan and any treatment notes.
- Surface faults known to be relevant to stability.

- Past opencast mining.
- Present opencast mining within 200 metres (although it is difficult to imagine a surveyor missing such a feature!)
- Potential future opencast mining within 800 metres.
- Existing subsidence claims, settled or unsettled; requests for preventive works; etc.
- For non-residential or development sites, certain legal information about withdrawal of support, etc.

For details you are recommended to check the latest position with the Coal Authority. Latest news at the time of writing is that areas in the Bideford-Chittlehampton area of North Devon must now have a coal mining search.

Quarries, etc

Most quarries are all too obvious (although they may have formed access points for underground works that are no longer apparent). However, they form convenient repositories for waste. Quite apart from the potential problems of subsoil stability, biological and chemical hazards may arise. Once a site has been totally filled, it is normally sealed and topsoiled, while giveaway methane ventilators may be required for many years. In theory, suitable raft foundations can be used to build successfully on closed landfill sites, but private purchasers in particular will want to be aware of any potential risks associated with the fill.

Study of old maps commonly reveals old quarries that are not shown on later maps. It is important to ensure that there are no underground workings from the base of such quarries.

Tunnels and other borings

As well as railway tunnels, there are other extensive man-made holes which are not for extractive purposes. Railway, canal and road tunnels (Fig.148) are normally unmissable, as they are usually shown on even 1:50,000 maps with clarity. However, some old disused tunnels may not be shown. In some cases, as with certain canals, it is now unclear whether the tunnels were even bored before the company failed, and in such cases further research may be advised. Tunnels and voids have also been created for military, sewerage, water extraction/supply and other purposes.

Many military sites re-used old mines, but others were entirely purpose-built and are, not surprisingly, little known although there are specialist clubs devoted to researching and documenting these sites (which are usually small). At Newhaven, Sussex (Fig.149), a modern housing estate stands over a series of shallow chalk tunnels used as a World War II communications centre. In North Wales, secret underground plants produced toxic chemicals.

The generally soft sandstones of the Permo-Triassic rocks have been especially used for tunnelling. Parts of Liverpool, Nottingham, and Bristol have been undermined thus for a variety of purposes. In the Smithdown Lane/Mason Street area, in the Edge Hill district of Liverpool, an extensive series of tunnels was created by a

Fig.148 (above) Some railway tunnels have a very thin cover: this one is estimated at less than 10 metres (note canopy of petrol filling station on road above).

Fig.149 (left) D-Day communications centre: housing estate constructed on top. The tunnels, driven in Chalk, are supported only by light steel framing and galvanised corrugated steel.

local eccentric, Joseph Williamson, for unknown purposes in the early 19th century. Williamson was excessively secretive. The full network has yet to be properly mapped, although part is already open to the public.

Schemes such as the Channel Tunnel Rail Link have given rise to a resurgence in rail tunnelling. At the time of writing, in Stratford, east London, an eight-metre boring machine was tunnelling some 20 metres below the surface when gardens in Lavender Street subsided to form a depression approximately 40 metres by 70 metres, and 10 metres deep.

Sewerage and water extraction headings are another little-known group of tunnels. In Bristol, Clifton Down is passed under by a major stormwater sewer.

Miscellaneous

Onshore oil extraction is not commonly thought of in the UK. However, there are scattered sites. Extraction of oil can cause subsidence under some conditions.

Wytch Farm, south of Poole Harbour, is possibly the best known. Here, extensive areas under Poole Harbour are pumped, using the latest extraction technology, from an onshore concealed site.

At Cootham, in West Sussex, the traditional 'Dallas'-style nodding donkey pumps are screened from the road a few metres away, so well that few passers-by can suspect them.

Old railway cuttings often 'disappear'. Older cuttings may not be properly filled, like old quarries, and may give rise to the same problems (Fig.150).

Fig.150 This distinctly sagging building is built on a poorly-filled railway cutting that formerly ran towards the viewpoint.

Landfill

Landfill sites (compare Quarries, above) are part of a specialist field of environmental geotechnical engineering. In the broadest terms, wastes can be divided into: municipal, medical, industrial, and hazardous.

Most municipal waste is disposed of in landfill sites. In best practice, municipal wastes are covered in daily in the landfill, either with soil, or with synthetic liners. Although most municipal waste is compacted in some way,

Cornish environmental problems

Cornwall, with its long history of metal mining (until 1998) has a particular problem. Much mine waste is acid: pyrite found as a common gangue mineral oxidises to produce sulphuric acid. Heavy metals are commonly found. Tin and arsenic predominate, but lead and zinc are also present. Metal pollution may be concentrated in old processing plant areas, or may have been dispersed by gravity and water, or by flue gases. Arsenic (which is found in the copper ore chalcopyrite) in particular has been spread in this last way.

Mine drainage waters also pose problems. In 1992 the recently-closed Wheal Jane issued water from a drainage adit that caused a 20-fold increase in Cadmium levels in the River Fal.

The long-term effects of such pollution on grazing, agricultural and allotment land are, so far, uncertain.

if only within the collecting vehicle, it still contains a high proportion of voids, and a high proportion of organic material which can rot. This means that considerable compaction can take place, while at the same time gases such as methane are generated by the decomposition of the included materials. Buildings constructed on such landfills should therefore have raft foundations. The gas is either vented off (now considered environmentally most unfavourable) or harvested for energy. Some developments have been constructed before gas venting has completely finished, but the tell-tale rotating cowls should give this away.

At Connon Bridge landfill site, in Cornwall, unstable refuse gave rise to a landslide in February, 2003. It was feared that the protective liner membrane might have been ruptured, with consequent risk to water supplies.

More hazardous materials are placed above special liners and may require suitably impermeable geology as a backup to resist escape of hazardous materials into groundwater, and hence the water supply. There are, however, many old tips that have been filled with noxious materials long before legislation began to control them, and unauthorized casual tipping continues.

Geophysical and other investigative techniques

Voids and other unsatisfactory footing conditions can be investigated by engineers in numerous ways. The following is a very brief summary.

- Vane test: extremely shallow, likely to work on subsoil only; measures resistance to rotating tool.
- Cone penetrometer: also shallow; measures resistance of material to penetration by conical-ended tool.
- Trial pits or trenches (only suitable for very shallow investigation).
- Boreholes: percussion, rotary coring (latter to unlimited depth in this context).
- Geophysical surveys: ground-penetrating radar; seismic; magnetic; gravity; resistivity. All these have different strengths and weaknesses, e.g. 20 metres would be best penetration for radar but shallow tunnels are likely to be shown clearly; a magnetic survey can be seriously disrupted by buried pipes, but is good at detecting old shafts.
- Sonar: in the case of the Edinburgh stone mine collapses, sonar equipment was lowered down large-diameter boreholes and used to explore the water-filled voids, just as a submarine senses its surroundings by sonar.

Chapter 12
Natural voids and related problems

Introduction

As well as man-made voids, there are plenty of natural problems. Limestone areas are most affected. The Carboniferous limestones of the Mendips, Wales, and parts of the Pennines have a large number of known cave systems, formed by slow dissolution of the limestone by rainwater. There is plenty of room for undiscovered caves, which do not have to be very large to cause severe damage to a building. The cover above such a void can be as little as a few metres. All areas of such limestone should be assumed to bear some degree of risk until the issue has been thoroughly checked. In the extreme, geophysical testing or boreholes may be necessary to confirm the situation. Fortunately, urban areas on these rocks are so far relatively limited in extent.

The Jurassic limestones of the 'Stone Belt' from Dorset to Lincolnshire are not noted for such cave systems. The Magnesian limestones of northern England do contain voids, although not to the same extent as the Carboniferous limestones. However, the voids in these rocks are usually due to the dissolution of former associated evaporites, which has caused collapse of the overlying strata.

The Chalk, which underlies so much of England, is not generally notorious for void formation, but swallow holes and cave systems (generally small) are known. The cliff face at Beachy Head intersects a long, tortuous passage which has been partially explored. The River Mole, where it passes through the North Downs, disappears into its bed in dry spells, only to reappear further on. Another well-known disappearing stream is the Mimmshall Brook, at the suitably-named Water End south of Hatfield. Recent events, however, suggest that the presence of old Chalk mines is a much more significant hazard.

Fig.151 shows the results of the collapse of part of a building on the Chalk and Head cliffs at Brighton in 1987. It had stood there for about 130 years before sudden catastrophic descent into an unsuspected void, so that longevity alone cannot be taken as proof of stability. The property has since been reconstructed in a style to match the original, following massive grouting of the void.

Fig.151 (above) Percival Terrace, Brighton: 1987 collapse.
Fig.152 (below) This sea cave in hard limestone extends under buildings but does not appear to be a problem. South Pembrokeshire.

Caves and voids-formation

In general, caves in the British Isles can be divided into Sea caves and Karstic caves.

Sea caves

Sea caves (Fig.152) are produced by the massive power of the sea eroding along pre-existing weaknesses in the rock. By definition, they are

Fig.153 This alarming cleft, nearly a metre wide, was produced by collapse into a sea cave beneath. South Pembrokeshire.

confined to the sea coast, and usually in areas of generally resistant rock. However, they may be of surprising extent. As they are mostly visible in the cliff, they are usually well known locally. Some caves in cliffs, however, are karstic and have been exposed by the retreat of the cliff.

As the available energy is so great it is not unusual for some sea caves to penetrate to the clifftop through holes (Fig.153), called blowholes or gloups, if the geological structure is favourable.

Karstic caves

Rainwater is naturally slightly acidic: it dissolves carbon dioxide from the atmosphere to form carbonic acid. This acid is weak, and has little effect on siliceous sediments. However, it is capable of dissolving calcareous sediments to form caves. Most caves are formed in indurated carbonate rocks because flow is concentrated along fissures (joints and bedding planes) which are then preferentially opened out to create chambers that may be of enormous size and form part of complex systems of anastomosing (separating and rejoining) passages. At the surface, hard limestones may form highly irregular pinnacled topography (described below).

The flow of groundwater in the Chalk and in the British Jurassic limestones is thought to be along fissures that are finer and more numerous, so that caves are rare in these rocks. (Chalk is often dissolved into pinnacled topography under overlying sediments, and forms a specialized subtype of karst - see below.)

The majority of significant British caves are in the generally hard lower Carboniferous limestones, although in Scotland, caves have developed in the Cambro-Ordovician limestones of the extreme north-west. Small areas of Devonian limestone in Devon also feature caves.

Most significant British caves (typical example shown in Figs.154 and 155) have probably been discovered, although their full extents are often unknown. The cave system with the highest vertical extent (over 300 metres) at the time of writing is Dan-yr-Ogof in South Wales. Gaping Ghyll, in Yorkshire, has a chamber 33 metres

Fig.154 (left) This cave entrance (view downstream) seems obvious enough. However, most of the river flow has already vanished into holes in the river bed further upstream.
Fig.155 (above) The (visible) resurgence: note the smoothly curved surfaces of a former phreatic state. Brecon Beacons National Park.

Caversham, Berkshire: Chalk karst subsidence

Caversham forms the hilly north-west part of Reading, north of the River Thames. Here, there has been an increasing number of recorded subsidence events causing structural damage. Reading-based engineers, Peter Brett Associates, have investigated many of the subsidence events in order to provide solutions to stabilise the affected properties. The low hills, which form spurs of the Chilterns, are of highly weathered Chalk which is overlaid by the Reading Formation. This comprises mostly sands in its lower beds and mostly clays in the upper beds. Gravel, probably ancient river deposits, caps some of the higher ground.

Geomorphological mapping was carried out by walkover survey and use was made of published geology and topographic maps. Drawing upon previous research including solution feature occurrence modelling using examination of aerial photographs and multispectral scanner imagery, it was concluded that solution features had formed at the Chalk interface with both the Reading Formation and the gravel. Solution feature formation probably took place in two cumulative phases where the gravel overlies the Reading Formation. Air-filled cavities can form both within solution pipe infills and within the Chalk itself.

A subsidence hazard map was devised in order to predict areas more prone to subsidence. The hazard mapping techniques have wide application to the Chalk outcrop. They can be used to determine areas prone to subsidence to assist planners, developers and insurers. Planners can then make the identification and mitigation of subsidence a condition of planning in line with PPG 14 *Development on unstable land.* See C. Edmonds, reference below.

This information is derived from EDMONDS, C. N. 2001. Subsidence hazards in Berkshire in areas underlain by chalk karst In GRIFFITHS, J. S. (Ed.) *Land Surface Evaluation for Engineering Practice.* Geological Society, London, Engineering Geology Special Publications, 18, 97-106.'

high (by contrast, at the time of writing the Gouffre Mirolda, in the French Alps, has a total explored system height of over 1,700 metres!).

Areas of limestone that are affected by such systems are called Karst, after an area in Yugoslavia. British karst is not so well developed as in the eponymous area, let alone the extreme 'Tower Karst' of China. There is a complex system of nomenclature. Surface holes that lead down into cave systems are termed swallow-holes, shakeholes, sinks, pots, etc, but such features are often rubble– or mud-choked and far from obvious. Sometimes they are blocked by glacial deposits and hold water (Fig.156). Caving clubs often deliberately dig out such entrances and gain access to previously inaccessible systems. The limestone surface in karst areas often exhibits characteristic solution features that may make walking progress very tiring.

Cave systems often reflect water levels of a previous time and entrances or exits may be submerged. Pressurised systems may reach below sea level (caves that are below the normal water table and are therefore always full of water are termed phreatic. Caves above the water table are vadose). Resurgences are usually more obvious.

Caves have often been found by miners' tunnels, so that caves and man-made voids cannot be assumed to be mutually exclusive. At Milwr Mine, in Flintshire, about 25 km of tunnels intersect numerous natural cavities including a lake that was used for the dumping of mine waste, thus creating an 'out-of-sight' hazard to water supplies.

In any such limestone area, you should assume that there is a high probability of undetected voids. These may be narrow, and therefore not be detected by drilling for site investigation; or they may be just too deep for geophysical detection. Pinnacled karst is especially difficult to investigate under later cover. The pressure for residential development means that more housing may be constructed on such areas in the future.

Fig.156 Pool in closed depression: a mostly-blocked swallow-hole (dark colour due to peaty water). Brecon Beacons National Park.

According to research at Nottingham Trent University, it is estimated that 90% of new sinkholes are caused by human activity changing the movement patterns of soil water and aquifers.

Evaporite dissolution

In north-east England, a belt of land running through Ripon and Darlington is affected by the appearance of sudden holes, including the water-filled group near Darlington known as 'Hell's Kettles'. Permian evaporites, notably the highly soluble gypsum, dissolve and remove support from the overlying ground, including later Triassic rocks. Because the dissolution of gypsum is much faster than in limestone, subsidence can be sudden and unexpected. At Ure Lodge, Ripon, a hole 6 metres deep was reported in 1997. Subsidence from this little-known cause cost £1.3 million in the ten years to 1999. As an aside, these features are said to have partly stimulated Lewis Carroll's creation of *Alice in Wonderland*.

Extensive areas of the Cheshire salt fields are also prone to natural subsidence of this type. Some of the pools called 'meres' that dot the county have originated naturally in this way, but many ('flashes') result from salt extraction (others are entirely unrelated and are post-glacial features).

Pinnacled bedrock and chalk pipes

Where carbonate rocks underlie permeable rocks, rainwater filtering through the overlying sediment can become (or remain) sufficiently acidic to dissolve the limestones, usually along routes determined by joint fissures. This can give rise to a highly irregular top surface for the limestone, into the hollows of which the overlying sediments slump, if they are relatively weak. This results in disturbed strata with inconsistent properties, and

Fig.157 Two solution pipes are clearly exposed here where the receding chalk cliff has intersected them. Note the sharp boundary with the overlying sandy and silty deposits. Height of visible cliff about 10 metres. There are dozens of pipes in this cliff section.

unpredictable limestone surfaces. Test boreholes (especially if widely spaced) can penetrate the pinnacles and miss the 'soft spots' in between, so that substantial unforeseen costs can be incurred when buildings or structures are erected.

In the Chalk (Figs.157, 158 and 159), such phenomena often give rise to 'pipes' of Cenozoic sediments that can penetrate many metres into the Chalk before they die out (as the acidic solutions penetrate into the alkaline Chalk, they lose their strength, so that the pipes tend to narrow downwards).

Where buildings are constructed over such features, they may suffer uneven support, leading to subsidence as the pipe fill compacts under additional load.

In the excavations for the North Downs Tunnel for Eurolink, pinnacled and piped Chalk was a particular hazard near the north (London) portal where some Chalk pipes penetrated over 80 metres down to the tunnel level (Warren and Mortimore, *Quarterly Journal of Engineering Geology and Hydrogeology*, Vol.36 Pt1, 2003) and contained Clay-with-Flints rather than Tertiary sediments.

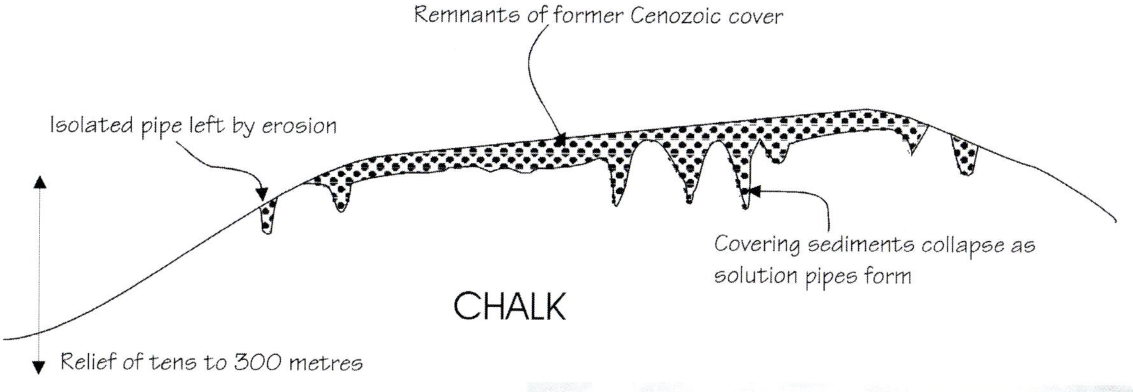

Fig.158 (above) Chalk pipe formation: schematic cross-section.

Fig.159 (right) Typical exposure of chalk pipes by earthworks for new road. Here the pipes (six are clearly visible) are full of reddish Clay-with-Flints.

Chapter 13 Flood risk and coastal recession

Types of coastline

Coastlines may be categorised as advancing or retreating.

Advancing coastlines occur where earth movements are lifting the crust, sea level is falling, or where lava, sediment or organic growth accumulates faster than the sea can erode it away.

There are minor areas of advancing coastline in western Britain where the crust is rising following the last Ice Age. The mass of the ice sheets was sufficient to depress the crust markedly; once it is removed, the crust rebounds (but very slowly), just as a lily leaf sags and rebounds as a moorhen passes over it. This is termed isostatic rebound.

Conversely, in many parts of Britain the coastline is advancing as river sediment is deposited. Many medieval ports in southern and eastern Britain, in particular, were forced to close as their estuarine approaches silted up. Boreholes show that great thicknesses of recent sediment underlie valley bottoms. However, the advancing coastline is often barely above sea level and can suffer reverses, especially in major storms when sand or gravel spits, or dune barriers, are breached. The whole system is dynamic, and seems static only on the puny timescale of a human lifetime.

Retreating coastlines occur as a result of downward movement of the crust, a rise in sea level, or erosion by the energy of the sea.

Much of the British coast, especially where formed in softer sediment, is visibly retreating on a timescale of years, or in extreme cases weeks.

Cliff retreat

Except where earth movements are raising them (or volcanic action forms new land - not normally a British problem), cliffs retreat under the action of the sea. Resistant rocks, such as the hard igneous and metamorphic rocks and older, indurated sediments forming much of northern and western Britain, often form cliffs that appear unchanged during a human lifetime or even longer.

Rather less resistant rocks, such as the Chalk, visibly retreat during such a time frame, and phenomena such as cliff falls at Beachy Head (Fig.160) are reported widely. Although the fallen mass of chalk inhibits erosion for some time, it is soon eroded away, and the process of collapse begins again. Unless such soft cliffs are armoured at their base by coastal defences, the rate of retreat may be in the order of a metre a year. Hence the terrace of Victorian cottages at Birling Gap, near Eastbourne (Fig.161, see overleaf), has already

Fig,160 Retreating coastline at Beachy Head, Sussex: Chalk can form vertical, or in this case slightly overhanging, cliffs despite being relatively soft. It fails by shearing along vertical joint planes.

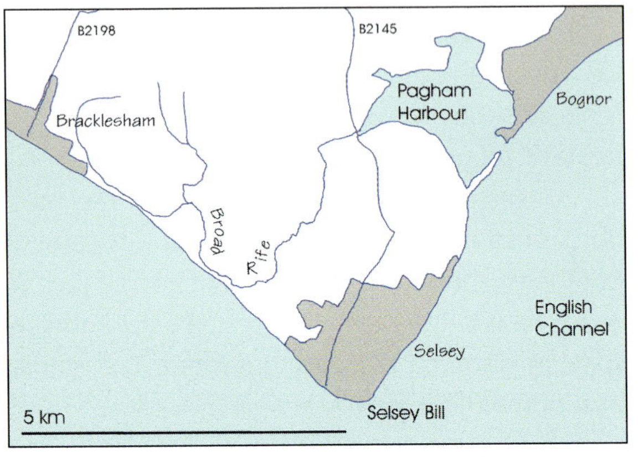

Fig.161 (top left) Birling Gap: truncated coastguard cottage terrace (and truncated wall in foreground). Fig.162 (top right) Sea-stacks near South Shields in retreating cliffs, mainly of dolomitic limestones. A natural bridge between the stacks fell only weeks before the picture was taken. Fig.163 (below left) Unconsolidated cliffs of Pleistocene silt and gravel rapidly eroding to the west of Selsey, West Sussex. Height of cliff above beach about 2.5 metres. This section of cliff retreated at an average 1.4 metres per annum during the 20th century. Fig.164 (above) Selsey is almost an island, with the Broad Rife valley barely 2 metres above mean sea level and less than 100 m from the retreating cliffs shown left.

been partly lost to the sea. Isolated more resistant sections (usually defined by joint patterns) may remain as stacks (Fig.162).

Softer or unconsolidated rocks (Figs.163 and 164) are easily washed away by the sea and retreat even more rapidly. The glacial deposits that form cliffs at Holderness, Yorkshire, have been the subject of television reports as farms fall into the sea.

Sometimes, cliff retreat may be slowed by the natural formation of protective pebble beaches, such as the famous Chesil Beach in Dorset, which absorb much of the sea's energy. Such protection depends on a continued supply of material from suitable rocks further along the coast (depending on the predominant direction of the sea's force). If this is interfered with, erosion and retreat can soon resume.

In the case of the village of Hallsands, in Devon, offshore dredging for gravel in 1897 completely altered the local marine energy budget. This led to lowering of beach levels, and in 1917 the village was destroyed by a storm.

Coastal and estuarine flooding and encroachment

Flooding is taken here to mean a high level of water that rises above a natural or artificial river bank, sea defence or normal coastal level.

In the UK, planning at the time of writing has involved scenarios in which the global mean sea level rises at rates between 2.4 and 10cm per decade.

Such rise is due largely to the effects of so-called 'global warming' whereby the world's ice caps and glaciers melt. The Greenland and Antarctic ice caps are so thick that if they were to be spread out evenly by complete melting, the rise in sea level would be catastrophic.

Much of Britain's coastal areas, and some inland areas such as the Fens and the Somerset levels, are very close to mean sea level (some parts of the Fens are below mean sea level). Study of Ordnance Survey maps will reveal spot heights of under 5 metres. At high tide, these areas are protected by existing sea defences. These defences are coming under gradually increasing pressure.

In the case of southern and especially eastern Britain, where slow subsidence is added to the rising sea level, the risk is correspondingly greater. While defences are maintained in good order, the risk is reduced, but considerable resources are needed. It is reasonable to suppose that while significant urban areas will continue to enjoy protection, some rural communities or isolated properties may one day be abandoned to their fate.

Water tables and aquifers

An aquifer is a body of rock that is significantly permeable by water and can hold such water in storage. Virtually all rocks can absorb some water into pore, cracks, joints, cave systems, or crystal structures (clay) but some, mostly clay or mudrocks, do not pass or yield significant quantities of water and are termed aquicludes.

A water table (Fig.165) is a notional surface that divides the zone of ground that is permanently saturated with rainwater at greater than atmospheric pressure (the phreatic zone) from that which is not saturated but through which water is percolating downwards from the surface (the vadose zone). Normally, where the water

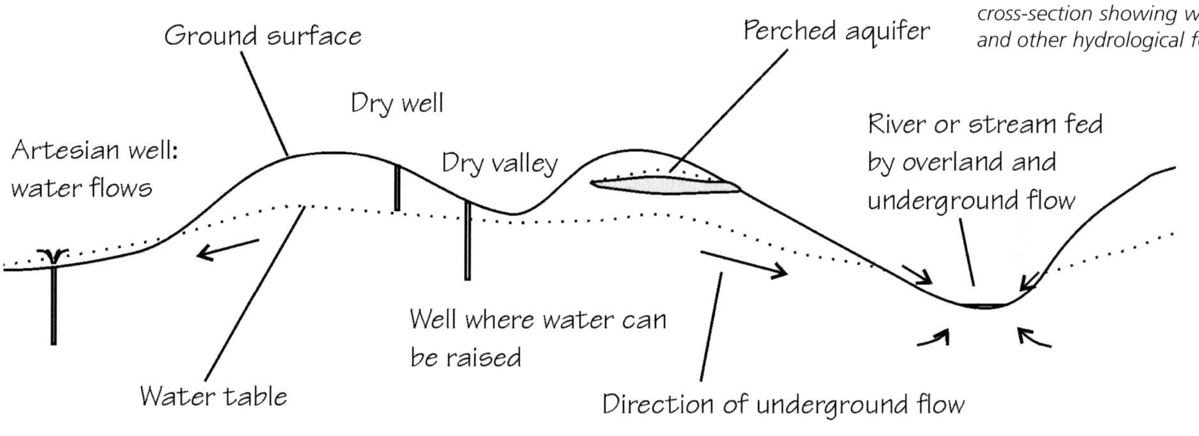

Fig.165 Hypothetical cross-section showing water table and other hydrological features.

table reaches the surface a stream or river flows. The water table is lowered where water is extracted by wells or pumping (such lowering is termed drawdown). Under a floodplain, the water table can be expected to be close to the surface, so that any basements or cellars will be naturally damp unless tanked. On hills, cellars should be drier. Historically, in central London the water table was greatly reduced well into the 20th century by pumped extraction. Reduction in such extraction, and changes in rainfall, have caused the water table to rise again.

In general the water table follows the shape of the ground surface, but with lower relief. Where it reaches the surface, water can flow, and may be concentrated at weaknesses such as joints, etc, to form a spring. Emergence may be diffuse and form a seepage.

Where a discontinuous aquiclude is present, a 'perched aquifer' can form, separated from other aquifers below. Wells sunk into such perched aquifers may be adequate for limited needs such as isolated dwellings. Hydrologists study the movement of water through the earth, and are employed by such bodies as the Environment Agency, water companies, etc.

River flooding

Atmospheric, seismic or technological hazards can lead to flooding; in general, however, persistent heavy rain is the usual cause.

The natural processes of river development generally lead to the river occupying a channel that follows a sinuous course which is confined within a more or less restricted, sub-level area called a floodplain. Most floods will be confined to this area, certainly in areas of moderate relief. Properties that are built on a floodplain are axiomatically at risk of flooding, whatever artificial precautions may have been taken.

In the long term, the river channel changes its path so that it sweeps in loops to pass over every part of the floodplain at some time or other. Rivers may deposit natural levees, banks that are raised above the floodplain level and are dry except in the most major floods, although they may be breached by water pressure and allow water onto the floodplain. At such times, buildings, etc, on undamaged levees may be safe.

The extent of the floodplain can easily be seen on most geological maps: it normally corresponds almost exactly with the area of alluvium laid down by the river, which is usually shown as a pale creamy-yellow tint. On a topographic map, except in areas of low relief it is also easy to note the extent of the floodplain when the vertical interval of the contours is 10 metres or closer. Construction of railway or motorway embankments, areas of made ground, etc, can confuse the issue, but not usually to any great degree.

Many factors contribute to the development of a flood. Apparently placid minor rivers can become roaring torrents, as did the Uck at Uckfield in 2000 (see box). In this case, particularly concentrated rain in a limited area seems to have been the cause. Areas underlain by clay that is already saturated may as well be tarmac in terms of runoff, so that rivers draining clay areas may suffer sudden massive increases in water level following only moderate rain (see boxes below and opposite).

Human activity

Floods can be modified or exacerbated by many human activities, many of which have been carried out without understanding the consequences:

- Urban areas with hard surfaces increase runoff.
- Smooth surfaces increase runoff rate.
- Embankments, undersized culverts, etc, trap water that could otherwise escape.
- Storm drainage or other diversion features may be undersized.
- Changes in agricultural use (e.g. unsuitable ploughing, land drainage, removal of tree cover) change runoff. A study at Essex University suggested that farming practices could account for a 14% greater risk of flooding.
- When artificial barriers or levees fail, the sudden rush can be catastrophic, and the effect can be prolonged because the water can only return to the river, when it falls, through the breach.

Areas below flood level may be drained by ditches that discharge through flap valves to the river at low water level, or may be lifted to river level by pumps (traditionally, in East Anglia, by watermills). Long-continued

Uckfield flooding

In October 2000, Uckfield in East Sussex suffered catastrophic flooding, graphic details of which were broadcast on national television.

It is less widely known that Uckfield was seriously flooded in January 1918, November 1929, January 1943, November 1952, November 1960 and November 1974. In the last case, a supermarket's stock was washed out of the shop, as in 2000.

The relevant points are:

- **The same month was involved in four of the six earlier occasions; and**
- **Floods seem to be easily forgotten.**

Chichester flooding

In late 1994, the River Lavant carried so much water following persistent heavy rain that much of east Chichester was flooded, including parts of the A27 trunk road. The River Lavant drains a large catchment in the South Downs (Fig.166), which debouches onto the coastal plain and feeds the bottleneck caused by culverts at Chichester. Moreover, the gradient between Chichester and the sea is so low that the river has changed its course many times in the past, forming a series of alluvial fan deposits over a wide area east of Chichester.

In dry years, much of the Lavant bed is dry and no problem occurs. When the Downs are saturated, however, heavy rain over the large catchment can result in a massive increase in runoff, leading in turn to bed avulsion (course-changing). In 1994, such a change of course might well have occurred but for human interference.

In 2000, disaster was averted by great effort involving prolonged fire brigade pumping. In 2002 a flood relief scheme was completed, which channels water (partly through tunnels) around a network of gravel pits east of the city towards Pagham Harbour and the sea.

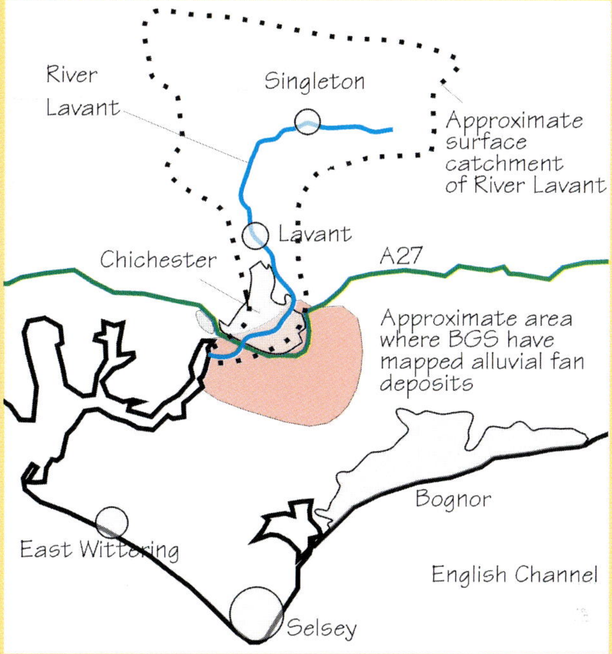

Fig.166 Chichester - catchment of Lavant.

usage may lead to a false sense of security that may be dispelled only by failure of the system. Levees may be raised artificially, but the river then deposits sediment in its own bed and eventually the process must be repeated to the point where it becomes highly hazardous. For instance, in parts of China's Yellow River the river bed itself (let alone the water level!) is 6-10 metres above the floodplain.

Statistics have been offered in recent years to estimate flood risk, leading to a state of mind in which terms such as 'hundred-year flood' have been used apparently without any great thought.

Following the floods in Britain in 2000 and 2002, there has been great outcry that 'they' should do something about it. This is an unrealistic posture. 'They' might as well seek to turn off the rain. Urban development over hundreds of years in unsafe places cannot be undone overnight, and flood defences cannot be relied upon: this year's flood defence is next decade's disregarded mound of earth. Quite simply, do not build on a floodplain.

A Sussex cottage

A client called me in to explain matters to his uncomprehending insurers. His ancient cottage had been extended recently. The planning authority had insisted that the roofline be unbroken, while at the same time Building Control had insisted that floor to ceiling heights be to modern standards. The agreed solution was to dig a hole and build the extension in it. This resulted in a narrow strip of land on the three outer sides of the extension, some 30-50 cm below natural ground level, at the foot of a gentle slope of clay. Surface water drainage from the lowered area was to the original well, by a pipe laid at a near-horizontal fall and prone to debris blockage.

In a wet autumn, the extension, but not the rest of the cottage, was flooded by runoff from the field behind.

No further comment is necessary.

Chapter 14 Reading the map

Mapping and its limitations

How the map is made

The field geologist

Despite advances in remote sensing, the production of a detailed geological map still depends on a booted and waterproof-clad geological surveyor with hammer, collecting bag and large-scale map (usually 1:10,000 in this country).

The geological map depends on the basic concept of a formation, that is a practical, mappable rock unit (not a time unit) that can be sometimes subdivided into recognisable 'members' or lumped together with other formations into a 'group'. As the surveyor of buildings is fundamentally interested in the performance of the rock rather than its date, this is quite satisfactory for most purposes.

The geologist will locate exposures, identify the lithology, look for fossils and collect samples if necessary for possible later laboratory examination. The attitude of the rocks will be noted, and suitable annotations made on the map, which is usually cut into small convenient sections termed 'field slips'. Cliffs, crags, quarries, road and railway cuttings and tunnels, foundation trenches (Fig.167), ditches and stream beds may provide adequate information but in south-eastern Britain natural rock exposures are generally sparse.

Fig.167 Foundered Tertiary strata exposed in temporary section on a construction site. The dark horizontal band is a modern soil horizon with made ground above. Yellow 1-metre rule to right. This exposure enabled me to make a correction to the BGS map.

Features will be located on the map by various methods. In developed western countries it is possible to plot directly onto excellent large-scale maps by simple surveying methods, but in remote areas compass bearings will be taken, or GPS equipment used.

Sometimes, boundaries may be plotted by reference to aerial photographs (see Remote Sensing).

In many cases there will be no exposures. In such event, the geologist may use a hand auger to penetrate the soil and obtain samples of the bedrock below. There may be well records in some cases, although these often require interpretation (or may be quite obscure) due to archaic descriptions of the rock penetrated. Field surface debris ('brash'), especially in ploughed land, may also be sufficient to allow the geologist to draw conclusions about the underlying rock. The trained eye of the geologist will assess minor features of the ground that may indicate changes in lithology and hence resistance to erosion.

The geologist will be careful to take into account the effects of soil creep, etc, which will result in the lower edge of brash features being downhill from the boundary of the relevant formation.

In many areas, the properties of the geology will change so rapidly over a short distance that it is unlikely that the surveyor in the field will be able to pick up the relevant features. For instance, in many parts of North Devon, there is a rapid alternation of dip which can only be inferred from the map by the apparent haphazard dips shown by the figured arrows. On the ground, the geological surveyor will have been confronted by a range

Fig.168 (top) Field geologists must be meticulous. The hammer marks the contact between dolerite on the left and baked mudstone on the right. At a glance they are very similar. Fig.169 (above) The figure is indicating the unconformity between purplish Precambrian sediments on the right and Ordovician sediments on the left. Both dip at the same angle (compare with Fig.90). Assiduous work is needed to avoid missing such features.

of fairly steep hillsides, mostly under pasture, with exposures limited to stream beds and the occasional quarry, etc. In such areas, small igneous intrusions could easily be missed unless they give rise to an obvious topographic feature (Fig.168).

Old and weathered exposures (Fig.169) require careful study to ensure that a vital feature is not overlooked.

Mine and tunnel records or new drivings may well provide extremely detailed information that can be used as a basis for working out the local stratigraphical sequence over a much larger area. Coalfields, in particular, have permitted very detailed understanding of the lateral variability of formations.

Dating and hence correlation of the rocks will be controlled in detail by laboratory investigation where necessary; in any case, any fossils collected will be examined by a suitable specialist.

Geophysical surveys of different types and deep research boreholes (to hundreds or even thousands of metres depth) supplement the field information and aid interpretation of deep structure. The concealed coalfield that underlies much of Oxfordshire and adjacent counties is known only from boreholes (solicitors are *not* required to make mining searches in this area!). The similarly concealed Kent coalfield was deduced to be likely in the 19th century, from general geological knowledge, and was confirmed by boring. There is an increasing tendency for Geological Survey maps to be compiled from sources that include fieldwork by university staff, sometimes under contract. Some maps include areas that have not been remapped since the mid-1800s. The Ludlow (181) 1:50,000 sheet, for instance, which was published in 2000, includes a sizeable area that was last mapped (at 1:63,360 scale) before 1855! Such areas usually show up on the map as apparently sparsely featured, with coarsely-plotted formation boundaries and no rock bodies shown at the Member level.

The remaining older maps must be used with some circumspection, because of the lack of information shown and the possibly misleading names of formations. In upland western Britain, in particular, huge stretches of country may be shown as one tint on the map, although they may comprise rapidly changing lithology that could have significantly differing performance.

Fortunately most urban and more densely built-up areas have been surveyed since the Second World War, although on some of the earlier post-war maps the keys need serious revision due to subsequent confusing changes in nomenclature. You should be wary of quoting the names of formations unless you are confident of their current names, or else indicate the uncertainty. Often, if two maps are placed edge to edge, the mutual boundary will show marked discrepancies.

The BGS has just completed digital mapping of the entire country at 1:50,000 scale but in general such mapping has been prepared by 'capture' of existing data, so that most of the same limitations of accuracy, etc, still apply.

Remote sensing

Remote sensing has been used extensively in mineral prospecting overseas, but has been of relatively limited use in Britain, with its often dense and/or artificial vegetation cover, and extensive built-up areas. It has been used, however, and a brief overview is appropriate.

Since the early 1950s, topographic maps have been produced to a large extent from high-quality aerial photographs. It was realised, even before the last world war, that careful study of these photographs would reveal geological features that were not immediately apparent in the field. Even features covered by soil are often apparent on aerial photographs (as indeed are archaeological features). Changes in vegetation cover can also reveal changes in subsoil and bedrock. This interpretation allows geologists to concentrate on key areas on the ground and to fill in the details later.

Airborne geophysical surveys permit the identification of local concentrations of high or low gravitational field or magnetic field, thus suggesting the areas where field geologists should begin to look for concentrations of economic minerals.

Since the 1970s, satellites in earth orbit have also been used for remote sensing. Satellite photographs are often taken in 'false colours', including beyond the optically visible spectrum. Computer manipulation of such images can reveal valuable detail. As an aside, the use of satellites can of course be useful for locational purposes to a field surveyor who is equipped with a GPS receiver. As is well known, the accuracy for civilian use is limited by the Pentagon, but in areas with few or no landmarks, the value to the surveyor carrying out reconnaissance can still be very great.

Formation colours and boundaries

On the 1:50,000 maps characteristic colours are used to show different formations/members. In general there is consistency of a sort: for instance, the Chalk formations are invariably shown by greenish shades, while the limestone formations of the Lower Carboniferous are in shades of blue. However, these tints often differ subtly across map boundaries. The colours are usually transparent enough to show at least the gross topography, but in more complex areas it can be very difficult to penetrate the colour and other ornament. A normal OS 1:50,000 map is useful to keep to hand, especially as the topography shown on the base map - especially the built-up areas - can be out of date by decades (the map margin shows the date of the base information).

Boundaries have traditionally been shown as unbroken lines where reasonably certain, pecked where uncertain, but the tendency on the most recent maps is to use unbroken lines for all solid rock formation boundaries, which seems a retrograde step.

A graphic key, often showing a generalised stratigraphical column, is provided and this usually (on modern maps) attempts to show variations in formation thickness, wedging-out, etc.

The graphic key may also provide useful lithological information. This is especially true of the more recent maps. Many maps will also provide one or more cross-sections which aid in interpretation of the structure. These are usually along carefully selected lines which will provide the maximum information.

Other information shown on the map sheet may include geophysical information (which is of no direct relevance to the surveyor) and a résumé of the geology.

You should bear in mind that small but quite important geological features may not be shown on the map because of the limitations of scale. There may be no clues to such a situation. Only knowledge of the geology will sometimes suggest local anomalies. In other cases, a body of rock may be relatively uniform on the large scale but may differ in properties on the small scale, for instance across the width of a building site. In such cases, unless there is knowledge of the nature of the rock this will not be apparent from the map.

Other features

The map may show old mine shafts or drifts/adits. These are shown conventionally, and are often only representative of the activity that has taken place in a particular area. Where a number of such features are shown close together, you should suspect the presence of many more. In coal mining areas in particular, areas of intensive past mining will often feature a history of outcrop mining, which is almost certainly not shown. This could result in uneven topography with a risk of small localised collapse, and backfill with uneven and unknown engineering qualities.

Finally, there are conventional symbols (Fig.170) for areas of worked-out coal or other minerals, made ground (i.e. landfill or tipping) and landslip. These are indicated by cross-hatching which can be confusing. Railway embankments, for instance, should be shown as made ground; while other areas of made ground are often found in harbour or industrial locations. Small areas will not be shown. Landslips, where shown, should suggest the presence of further landslips which are too small to be depicted. There will also be landslips which have not been detected, and have been shown as some other form of superficial deposit. Such landslips are often only recognised when excavations are cut into them.

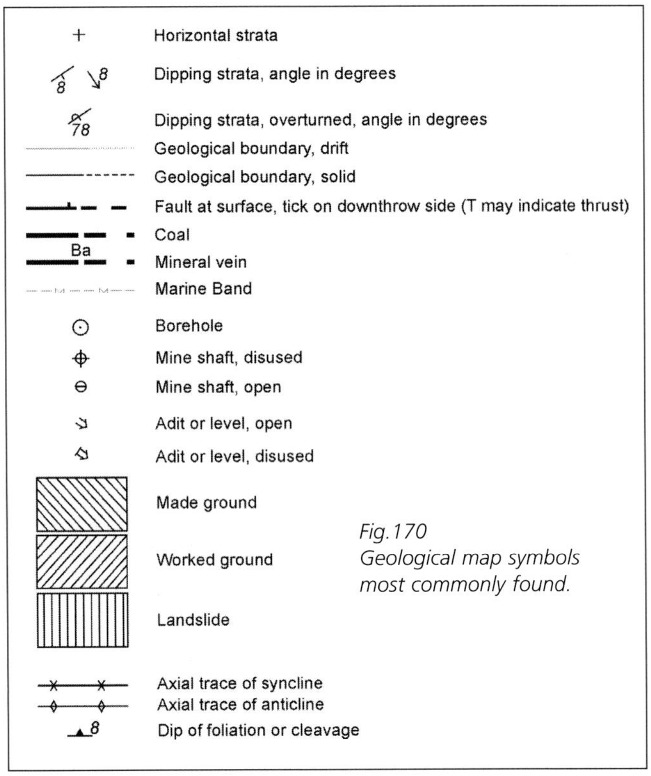

Fig.170
Geological map symbols most commonly found.

Solid (bedrock), drift (superficial deposits), and combined map types

You should take care to distinguish between the superficial deposits (so-called 'drift') and other deposits. Although superficial deposits are treated as relatively insignificant by many geologists, they form the subsoil to much of the built environment and, in surveying terms, are therefore as important as more 'glamorous' rocks such as Portland stone or materials containing exciting fossils. As drift deposits are often extremely complex, it is very difficult for them to be depicted adequately on the map. In almost all such cases, some sort of trial excavation would be necessary to establish the properties of the subsoil although it is often possible to make deductions from the nature of the underlying geology. Except in the case of alluvium, much drift has only travelled short distances and is likely to be a structureless mixture.

'Solid' geological maps do not show the superficial deposits, although some may show the outlines without any colour fill. This could lead to unintentional confusion, especially for the inexperienced reader. 'Drift' maps show the full extent of drift information available at the time of the survey. Some maps may be described as 'solid and drift' or 'solid with drift', which is perhaps more meaningful as nearly all modern maps show the solid geology by outline underneath the drift. Only old drift maps fail to show the solid outlines. Especially in coal mining areas, knowledge of the geology below the drift can be very important. The properties of the drift, in such cases, will be almost irrelevant if there are (say) old mining galleries only a few tens of metres below the base of the drift.

The presence of drift, of course, means that the outlines of the solid geology beneath are usually inferred, by means of extrapolation supplemented by wells, bores, and pits etc. As the map will not necessarily show the data constraints, further research may be necessary.

Folds, faults, dips, etc

The structure of the rocks is shown in several ways (Figs.170 and 171). Dips measured in the field are shown by arrow or T symbols, with a figure denoting the dip angle. Overturned strata (i.e. those that have been moved so that they are upside down from their original depositional attitude) have special symbols. Sometimes anticlinal and synclinal axes are obvious enough to be mapped and indicated directly.

Most significant faults are not directly observable, but are inferred from the field relationships of the relevant formations. They are usually shown by thick lines. Mineral veins look like faults on most maps and can be hard to distinguish unless there is a symbol denoting the material extracted (for instance, Cu next to a vein denotes copper ore). It may help if you try to visualise the pile of strata as a multilayer sandwich with a rather uneven filling.

Map symbols

The Geological Survey necessarily uses consistent symbols, which are generally echoed by other mapmakers. There are slight variations, however, depending on the draughting fashion at the time of production.

The symbols may be very small and can be hard to pick out against a dark background, so that careful scrutiny is advised.

The dummy map (Fig. 171)

The simple geological map provided here attempts to demonstrate the basic principles of map interpretation, including assessment of structure and reconstruction of history and environment. Although the map is extremely simple, you should note the following points in particular:

- The anticlinal inlier in the stream valley.
- The pinching-out of the conglomerate that rests directly on the granite.
- The unconformity beneath the unfossiliferous sandstone (outlier clearly cross-cutting earlier formation boundary).
- The outlier of oölitic limestone.
- The 'V' shape formed by the stream valley: the V always points in the direction of dip.

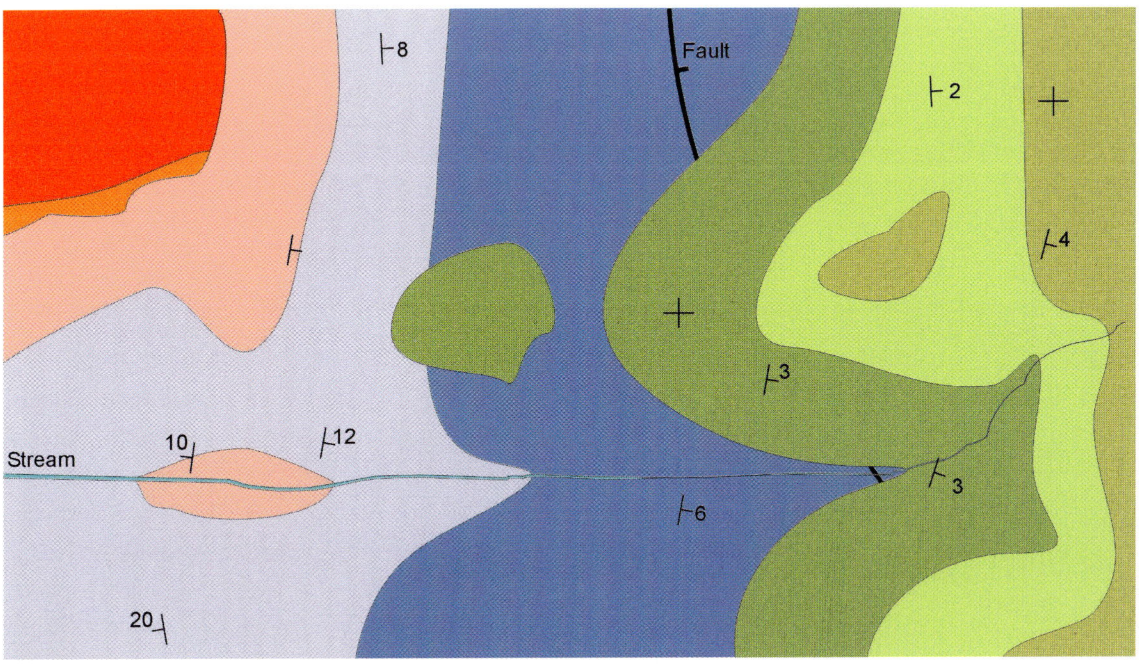

Oolitic limestone with ammonites

Mudstone with ammonites

Sandstone with no fossils

Limestone with clams and ammonites

Mudstone with oysters

Sandstone with dinosaur bones

Conglomerate

Granite

Interpretation based solely on map and key:

A granite in the north-west of the area was eroded, probably subaerially, and partly mantled in conglomerate before terrestrial sandstones were deposited during the Triassic period [*terrestrial fossils*].

A marine transgression [*oysters present]* occurred, after which there was minor folding, faulting and uplift [*steepish dips, anticline exposed in stream valley*]. This was followed by renewed deposition of unfossiliferous sandstone [*sandstone unconformable on earlier rocks and structures*], then by a marine transgression [*ammonites present*] during the later Mesozoic. The sea floor shallowed later and oolites were deposited under warm-water conditions.

IMPORTANT NOTE

At the time of going to press the BGS has just issued a statement that on most new maps and some reprints of older maps, the terms 'drift' and 'solid' will be discontinued and replaced by 'superficial deposits' and 'bedrock' respectively. As there are large stocks of existing maps this transition may, however, be slow in practice.

Appendix 1 Further reading and reference

For those who wish to take further interest in geology (and who will rapidly find just how much I have had to generalise!), or to have works of reference on particular areas, a variety of publications are available.

Basic textbooks

Holmes' Principles of Physical Geology
P. McL. D. Duff
Chapman & Hall 1993
is a useful general starter, but is a thick book and takes some wading through for the uninitiated.

Unlocking the Stratigraphical Record
Ed. P. Doyle and M. R. Bennett
John Wiley & Sons 1998

Introduction to Geological Structures and Maps
G. M. Bennison & K. A. Moseley
Arnold 1997

The Geology of Britain: an Introduction
P. Toghill
Airlife Publishing 1999

The above authors list further sources.

By way of interest...

... rather than dry text:

Richard Fortey (based at the Natural History Museum) has written a number of very accessible books such as ***Trilobite*** and ***Life: An Unauthorised Biography*** although these are mostly of background interest.

Simon Winchester's book ***The Map that Changed the World*** is about William Smith, canal surveyor extraordinary, who drew the first geological map of England.

Local reference

In the UK the main source is, of course, the 1:50,000 scale coloured maps of the British Geological Survey (BGS). These can be obtained from better bookshops in some cases (normally only local maps will be carried), or conveniently purchased online (www.bgs.ac.uk at the time of writing). You need to know which sheet is relevant (the sheet boundaries follow archaic Ordnance Survey map boundaries), so a key map is necessary. Some areas have 1:25,000 maps. 1:10,000 maps can be purchased for many areas, and a range of more specialised maps (such as gravity anomalies) are also produced. Contact BGS as above, or at:
British Geological Survey, (BGS Main Sales Desk – All Mail Order sales plus BGS Shop)
Keyworth, Nottingham NG12 5GG.
Tel: +44 (0)115-936-3241 Fax: +44 (0)115-936-3488
e-mail: sales@bgs.ac.uk

The *British Regional Geologies* provide an overview of geology in the following areas, but a few volumes are very much out of date.
Orkney and Shetland
The Northern Highlands
The Tertiary Volcanic Districts
The Grampian Highlands
The Midland Valley
The South of Scotland
Northern England
The Pennines and Adjacent Areas
Eastern England from the Tees to the Wash
Central England
The Welsh Borderland
East Anglia and Adjoining Areas
London and the Thames Valley
The Wealden District
The Hampshire basin and Adjoining Areas
Bristol and Gloucester District
South-west England
North Wales
South Wales
Regional Geology of Northern Ireland (not uniform)

Many map sheets have explanatory memoirs, although again these are often out of date and require some knowledge to update them. However, they usually provide detailed information of exposures (which have often since vanished) to supplement the map. The memoirs are rather expensive, but those local to your area are well worth the money: they will stay on the shelf for years.

The BGS provide GeoReports (a new service in January 2003) which can be ordered online (www.bgs.ac.uk). However these are naturally based on such evidence as is available (indeed the document contains a number of disclaimers), so that you may still not be satisfied. However, such a document can still be very useful and is quicker to obtain than purchasing a geological map.

The Coal Authority's Mining Report Service (www.coalminingreports.co.uk) also provides a useful service (used by Solicitors who search in mining areas). Contact them at:
The Coal Authority
200 Lichfield Lane, Mansfield
Nottinghamshire, NG18 4RG
01623 427162 or thecoalauthority@coal.gov.uk

The *Geologists' Association* publishes field guides to certain areas which provide some general information and can be used as self-help if you wish to mount your own field trips. Contact them at:
The Geologists' Association
Burlington House, Piccadilly, London W1V 9AG
02074349298 or geol.assoc@btinternet.com

In The Irish Republic the main source is The Geological Survey of Ireland (GSI), whose most useful map scale is 1:100,000 although a few 1:25,000 maps are also published. As with the BGS, a range of specialised maps are on offer. The 1:100,000 maps come with an explanatory booklet with both lay introductions and technical expositions.
Contact GSI at gsisales@gsi.ie
or Tel: 353-1-678-2868 Fax: 353-1-678-2549

If you wish to attempt fossil identification, suitable guides include:

Fossils of the Chalk
Ed A. B. Smith and D. J. Batten, The Palaeontological Association, 2003, probably the most useful publication from this organisation

British Palaeozoic Fossils/British Mesozoic Fossils/British Cenozoic Fossils
British Museum (Natural History), various dates.

Special reference

There are too many works on specific issues for many to be singled out. However, you may find the following useful:

Health Effects of Climate Change
Department of Health
available in hard copy from: Air and Noise Pollution Unit, Room 693D, Skipton House, 80 London Road, Elephant and Castle, London, SE1 6LH;
also available online at time of writing:
www.doh.gov.uk/airpollution/climatechange02/): much detail on flood risk.

PPG 14 issued by the Office of the UK Deputy Prime Minister is a neat summary of subsidence risks.

English Stone Building
Alec Clifton-Taylor and A.J. Ireson
Victor Gollancz 1994
The Environment Agency website at time of writing (www.environment-agency.gov.uk) can be useful for flood information, but the maps require careful interpretation.

Appendix 2 Glossary

Note: Many of the terms explained are found in this book but some are additional. Inevitably I have had to be selective, but the numerous mining terms will assist in interpreting old documents, etc. I have concentrated especially on those terms that may be found on maps or define various sorts of void, for obvious reasons.

Note: usage varies and/or overlaps between authors, mining areas, etc. *Italics* within definitions refer to cross-references.

Acid
Adjective for *igneous* rocks with more than 10% free silica.

Adit
Sub-horizontal mining tunnel driven into hillside, usually self-draining or entirely for drainage purposes (for drainage, also *water gate, sough*).

Age
Span of geological time smaller than an *epoch*.

Agglomerate
Volcaniclastic sediment formed of *bombs*.

Alluvium
Unconsolidated silt, clay, mud or gravel deposited by rivers or streams; usually short-lived in geological terms in any area subject to erosion, but can ultimately accumulate in subsiding basins to form immense thicknesses of sediment.

Alpine
1. Of the Alps.
2. Term given to period of major earth movements (*orogeny*) that gave rise to the main mass of the present Alps and to some significant movement in southern Britain; Eocene to Miocene in age.

Amphibole
Group of important, usually dark-coloured, rock-forming minerals in igneous and metamorphic rocks; crystalline structure characterised by double chains of silicon tetrahedra and sodium, calcium, magnesium and other metal ions.

Amphibolite
Dark rock, often with obvious elongated texture, formed by intense *regional metamorphism* of basic igneous rocks; dominated by *amphibole* minerals.

Andesite
Common intermediate lava, usually erupted in volcanic chains next to a subducting trench (named from the Andes of S. America) while *granite* forms deep below.

Anticline
Upfold in sedimentary strata; cf. *antiform*.

Antiform
Looks like an *anticline* but may include inverted strata.

Aquiclude
In effect, the opposite of *aquifer*.

Aquifer
Body of rock permeable by water (may be capable of exploitation for water supplies).

Argillaceous
Rock containing clay- or silt-sized particles; shales, mudstones and clays. Argillaceous limestone, limestone with significant clay content.

Argillite
Indurated *mudstone*.

Ash
Fine-grained *pyroclastic* material.

Avalonia
Land mass that now forms one of the major *terranes* of the British Isles.

Barite *(Caulk, Calk, Cawk)*
Barium sulphate, pale-coloured and unusually dense and usually found in veins.

Basalt *Dhustone, Whinstone*
Basic eruptive lava, often emitted at ocean ridges (e.g. Iceland). Very common, with various distinctive varieties; e.g. tholeiite, which contains quartz, forms plateau lavas and is erupted at mid-ocean ridges.

Basement
A relative term:
1. Rocks below the rocks that are under discussion or of interest.
2. Generally old, deformed, metamorphosed or igneous rocks that lie below the rocks which are under discussion or of interest.

Basic
Adjective for *igneous* rock with 45-55% silica, typically with no quartz, and with calcic *felspars* and with *pyroxene*.

Basin
1. Depression in crust, usually of significant extent, that accumulates sediments.
2. Used to describe short downfold or *pericline* that leaves subcircular or oval outcrop pattern on geological map.
Batholith
Originally deep-seated, large, mass of *igneous* rock, commonly formed beneath rising mountain chains; *granite* and its relatives are the most common rocks. The Dartmoor granite is the best-known British example.
Bed
Subdivision of *formation* or *member*.
Bedding
Horizontal layering of sediments, coarser scale than *lamination*.
Bedding plane
Distinct surface separating two beds. *Dip* measures the angle of bedding planes.
Bedform
Features such as ripples, dunes etc at the interface between the depositing fluid (air or water) and existing sediment.
Bedrock
1. Unweathered rock of all types beneath *regolith*.
2. In Ireland, term for what in UK is termed 'solid'.
Bell pit
Obsolete type of shallow mine with narrow vertical entrance, usually under 12-15 metres deep, wider below. Usually found in groups near coal seam or ironstone outcrops.
Bentonite
Montmorillonite-rich clay (see *clay minerals*) formed by alteration of volcanic ash and tuff; used for its absorptive properties in industrial processes, including clarifying liquids in the food industry. Cf. *fuller's earth*.
BGS
British Geological Survey, government agency that makes geological maps of the UK at various scales, of which 1:50,000 is the most useful; also publish reports for each map area, called Memoirs or (briefer) Sheet Explanations.
Bioturbation
Churning action of organisms on sediment that results in loss of original depositional features and may introduce biotic elements e.g. faecal pellets; may produce burrows or other traces that can be preserved.
Bivalve
Clam; lamellibranch (obsolete), pelycypod; mostly marine Mollusc class with muscular foot, like cockles and their relatives. Found in rocks from Cambrian to present.
Black Jack
Sphalerite.
Blackband ironstone
Carbonaceous, dark-coloured version of *clay ironstone*.
Boghead coal
Type of *coal*.
Bombs
Material ejected from volcano while still molten, over 64 mm diameter.
Bone bed
Concentration of vertebrate bones formed by removal ('winnowing') of fine sediment.
Bord and pillar   Pillar and stall.
Boulder Clay
Clay with boulders, pebbles, stones etc, produced by glacial action; obsolescent term for material that is usually a type of *till* but could be from some other process, such as solifluction or debris flow.
Brachiopod
Marine bivalved organism, important until end Cretaceous; now minor.
Breccia
Very coarse-grained sedimentary rock formed of angular *clasts*; cf. *conglomerate*.
Calamine
Smithsonite, zinc carbonate, minor ore of zinc.
Calcareous
Limy; containing calcium carbonate.
Calcite
Most common carbonate mineral, mostly calcium carbonate; major constituent of limestones (and many shells); can cement sandstones.
Caledonian
Term given to period of major earth movements (*orogeny*) that gave rise to the massive mountain ranges in what is now northern and western Britain (also parts of present Scandinavia and north America); associated with final closure of Iapetus in late Silurian to Devonian times, but started much earlier.
Calk
Barite.

Camber
Dip of hard strata into valley, caused by plastic flow of underlying softer strata rather than folding; a local not a tectonic problem; associated with *gulls*. In some cases, associated with mining that has removed support.
Cannel coal
Type of *coal*.
Cap rock
Impermeable rock that traps oil or gas below.
Caulk
Barite.
Cawk
Barite.
Cenozoic
Era of Phanerozoic including Palaeocene, Eocene, Oligocene, Miocene, Pliocene and Pleistocene periods and systems (or sub-periods and subsystems, according to usage); see illustration of Geological Timescale.
Cephalopod
'Head-foot': octopuses and their relatives; mostly shell-bearing. Cambrian to recent; some groups, such as ammonites (Triassic to Cretaceous), useful for dating.
Chalcopyrite
Copper iron sulphide, ore of copper, often in veins.
Chalk
White or grey limestone, relatively soft, widespread in southern and eastern Britain (and in France, parts of USA); formed from aggregate of calcareous microorganisms. Often contains *flint*.
Chert
Dark variety of cryptocrystalline silica, often forming nodules or less regular masses, in sedimentary rock, often black shales with submarine lavas nearby; cf. *flint*.
Clast
Literally, broken fragment; sedimentary rock particle.
Clay
Sedimentary particles of finest size.
Clay ironstone
Beds or nodules of *siderite*-impregnated mudrock, often forming workable ore, also termed Mine in some areas.
Clay minerals
Loose term used to describe mainly minerals with small platy crystals (hydrous aluminium silicates), often with capacity to swell or shrink according to adsorbed water content (as in the dreaded Shrinkable Clay); may be plastic in mass when wet. When baked (or fired), often mixed with sand, etc, yield refractory products – bricks and tiles – or pottery. Includes Montmorillonite and other Smectites, Illite and Kaolinite etc.
Clay-with-Flints
Deposit found on chalk, comprising mixture of solution residue from the chalk, uneroded flints and variable proportions of Tertiary strata churned up by solution subsidence and glacial or periglacial frost action.
Cleavage
1. Preferred fracture surface arrangement of mineral crystal, related to atomic structure.
2. Rearrangement of (especially) fine-grained rock by pressure, producing a more or less fissile material that splits at right angles to the original pressure direction; slate has a strong cleavage.
Climbing shaft
Shaft.
Coal
Solid sedimentary deposit formed from carbonised organic matter, extracted and used as fuel because of its high carbon content; see appropriate section of book.
Coal Bed Methane (CBM)
Methane extracted from unworked coal seams.
Coal Measures
Loose term indicating strata that contain coal seams; in UK, the Westphalian or upper part of the Silesian (i.e. Upper Carboniferous).
Coal Measure Methane (CMM)
Methane harvested from coal mines for fuel purposes.
Collar, shaft collar
Section of mine shaft lining which projects above ground.
Colluvium
Weathered or eroded rock debris that has been moved down the land surface by creep or by surface wash.
Compression, compressional
Used to refer to sections of the crust which are being compressed laterally.
Conformable
Succession of sedimentary strata without any appreciable break in deposition or *unconformity*.
Conglomerate
Very coarse-grained sedimentary rock formed of smooth *clasts*; cf. breccia.

Cornstone
Concretionary limestone, a palaeosol, commonly found within the *Old Red Sandstone*.
Creep
Slow downhill movement of earth material caused by freeze-thaw processes, rainwater, biotic activity etc; can overturn boundary walls, poles etc.
Cross-bedding, cross-stratification
Arrangement of sedimentary bedding with curved and interrupted bedding planes that indicate formation by various types of sandwaves, dunes, or ripples etc; indicative of formational environment; common in exposures.
Crosscut
Sub-horizontal mining tunnel; cf. *level, drivage, adit, heading*.
Crosscut
Mining tunnel driven across major trend of the structure, e.g. inclined bedding or coal seams, veins, etc.
Cross-stratification
Common sedimentary structure formed by dune or ripple migration; typical of rivers, deltas, offshore bars etc, and sand dunes.
Crown hole
Hole at surface produced by progressive collapse of mine workings, forming a void which works its way to the surface as collapse progresses.
Cryoturbation
Process of churning up of bedding and other depositional features by the action of ice; typical of near-glacial environments. Also termed geliturbation.
Crystal
Structured organisation with regular atomic lattice assumed by many solid materials as they form; may be bounded by regular crystalline faces with distinctive arrangement where conditions permit.
Crystalline limestone
Limestone that has recrystallised partially or extensively, or is of crystalline grains with crystalline cement.
Crystalline rock
Loose term for most igneous and many metamorphic rocks, formed of interlocking mineral crystals; also *crystalline limestone*.
Deads
Metal mining term for useless or waste material, often piled up in old workings. Cf. *goaf*.
Dextral
Of a strike-slip fault: movement of the far side of the fault to the right of the viewer.
Dhustone
Basalt.
Diachronous, diachronic
A lithological unit that differs in age from one location to another.
Diagenesis
Processes that act on sediment after deposition, including compaction, cementation, nodule formation etc.
Diamicton
Sediment with sand or coarser material supported in matrix; diamictite, lithified version of such. Non-committal term without inference as to cause/source. Includes *till, tillite*.
Diapir
Mass of rock (usually relatively soft and mobile, like salt) that penetrates overlying strata; adjective diapiric.
Diorite
Deep-seated equivalent of *andesite*.
Dip
Angle at which sedimentary beds slope from the horizontal; measured off *bedding plane* using a clinometer.
Dolerite
Equivalent of basalt, usually found as dykes and sills; finer than Gabbro, but may contain visible crystals.
Draw, angle of
Angle between vertical and the notional plane that runs from a coalface (or similar) to the outermost line of surface subsidence; the surface effects are in advance of the coal face.
Drawdown
Reduction of water level in well or borehole, usually expressed as so much for a given rate of pumping, etc.
Drift
Sub-horizontal mining tunnel; or tunnel driven into ground in plane of coal seam. See also *Superficial deposits*.
Drive, drivage
Sub-horizontal tunnel to, through or near ore-body; cf. *roadway*; drive (verb), to create such tunnel.
Dyke
Vertical to sub-vertical igneous intrusion, more or less planar; may be associated with and/or radiate from volcanic centre. Sometimes very persistent in length, such as the Tertiary Acklington Dyke of northern England.
Engine shaft
Shaft.

Epoch
Span of geological time smaller than a period.
Era
Major division of Geological Time.
Esker
Linear, mainly sandy body, with cross-stratification. Deposited by glacial meltwater, probably in sub-glacial or englacial tunnel.
Evaporite
Rock formed by evaporation of water containing salts that are then precipitated.
Exposure
Area of rock visible at the surface, without soil or vegetation, in crags, cliffs, stream banks etc; as in "the formation is exposed at …" cf. *outcrop*.
Extension, extensional
Used to refer to sections of crust which are being stretched apart laterally, probably with a line of spreading.
Face
Surface of quarry or mine where extraction is taking place.
Facies
The totality of features of a sedimentary rock, such as *lithology*, fossil fauna, *sedimentary structure*, etc. which, together, provide information about the depositional environment etc.
Fault (normal fault)
Approximately plane fracture surface caused by brittle fracture, with observable displacement; result of tension.
Felspar, feldspar
Rock-forming silicate mineral, with highly variable composition; varieties are sodium-rich (sodic) or potassium-rich (potassic), found in more *acid* rocks; or calcium-rich (calcic), found in more *basic* rocks.
Fireclay
Seatearth.
Firedamp
Methane.
Flat
Horizontal to sub-horizontal mineral vein, laterally extensive. Northern British use.
Flint
Variety of *chert* (very finely crystalline silica) found in *chalk*, giving distinctive appearance to chalk cliffs; usually burrow infill or shell replacement.
Flood
Overtopping of normal levels, or of natural or artificial barriers, by water.
Floodplain
Area of land adjoining a watercourse that floods naturally, often every winter.
Floor
In coal mining, rock below seam.
Fold
Bend in rock strata or other planar feature. Bewildering complexity of sub-terms.
Formation
Fundamental unit of lithostratigraphy; a mappable rock unit. Several formations can form a *group*.
Fossil
Preserved remains of an organism or its traces.
Fossiliferous
Bearing fossils; cf. *unfossiliferous*.
Fuller's earth
Similar to *Bentonite* in properties, but less pure. Originally used by fullers to remove grease from wool. Also, confusingly, the Fuller's Earth Formation of the British Jurassic, with fuller's earth deposits that are no longer worked.
Gabbro
Deep-seated equivalent of basalt; often faintly or strongly greenish.
Gabion
Heavy-duty wire mesh basket-like container, filled with rocks. Used as retaining wall for landslides, etc and to add weight to landslide toes, etc.
The construction offers no barrier to water.
Gallery
Sub-horizontal mining tunnel.
Gangue
That part of an ore vein that is valueless and is discarded.
Ganister
Siliceous *palaeosol* developed under coal.
Cf. *seatearth*.
Gastropod
"Stomach-foot": snails and their relatives; Cambrian to recent, mostly marine.
Geliturbation
Cryoturbation.
GSI
Geological Survey of Ireland, state agency that prepares Geological maps at various scales of which the most widely available is 1:100,000, and explanatory booklets.
Ginging
Mine shaft lining of masonry.

Glaciofluvial sediments
Produced by the action of running water associated with glaciers, on, within, under or downstream from the glacier.

Gneiss
Coarse-grained banded rocks with typical texture, formed during high-grade *regional metamorphism*, often from deep-seated igneous rocks.

Goaf
Waste used to backfill exhausted sections of mine (usually coal). cf. *deads*.

Gob
Goaf.

Gondwana, "Gondwanaland"
Huge southern-hemisphere continent of late Palaeozoic times; *Avalonia* split away from its northern edge.

GPR
Ground-penetrating radar, used to detect shallow voids.

Granite
Deep-seated equivalent of Rhyolite; usually distinct crystals (*macrocrystalline*); dominant in most large intrusions; good building stone; usually found as quite large bodies, *batholiths*, in eroded mountain chains.

Granitisation
Conversion of existing rocks to granite at great depth and high temperature.

Graptolite
Lower Palaeozoic colonial organism, usually found as flattened film in dark shales, etc; important for correlation of strata.

Group
Lumping together of adjacent *formations* with some commonality.

Gull
Cleft in rock caused by opening of joints in hard beds associated with Camber.

Hade
Obsolete term applied to faults and mineral veins; denotes angle from vertical (cf. *dip*).

Head
Type of *colluvium* mostly thought to be relict of near-glacial weathering.

Headgear
Structure supporting winding sheaves of mine shaft cages.

Heading
Sub-horizontal mining tunnel; cf. *level, adit, drivage, crosscut*.

Horizon
1. Distinguishable layer in soil.
2. Vague term used by stratigraphers to denote a particular bedding plane or relatively thin bed, as in 'the horizon of the subcrenatum marine band'.

Hornfels
Metamorphic rock formed next to igneous intrusion, with splintery fracture.

Hushing
Method of exposing mineral veins, etc by impounding water and then allowing it to wash soil, etc from hillside.

Hypabyssal
Igneous rocks formed at fairly shallow depth; cf. *plutonic*.

Iapetus
Name given to a sea or ocean that formerly separated Laurentia and Avalonia, the two main *terrane* groups that form the present British Isles.

Igneous
Rock that has solidified from molten or partly molten material.

Ignimbrite
Internally welded pyroclastic-flow (*volcaniclastic*) deposit.

Incline
Sloping tramway for waste or mineral haulage; typically, dead straight in plan and at constant angle. Often still obvious on map (often running from position next to present or former railway) or on site, can provide valuable clue to former mining activity.

Induration
Processes that lead to the hardening of a rock (or other material).

In situ
Found in position in undisturbed rock or location and therefore providing more information about that rock or location.

Inlier
Small area of older rock completely surrounded by younger rock; cf. *outlier*.

Interfluve
High ground between adjacent river valleys.

Intrusion, intrusive
Intrusive rocks are igneous rocks that formed when magma penetrated upwards into the crust from magma-generating locations; an intrusion is a particular body of such rock.

Ironstone
Sediment with high iron concentration; cf. *clay ironstone*.

Isotope
Variety of an element that has a different number of neutrons in its nucleus. For instance, Uranium-235 (U-235) has 235 neutrons, while U-238 has 238 neutrons. Carbon-14 dating (radiocarbon) of archaeological remains, for instance, depends on the decay of the very small quantities of radioactive C-14 which is mixed with the C-12 contained in organic compounds.

Joint
A fracture (usually more or less at right angles to sedimentary bedding) with no observed displacement.

Kettle hole
Hollow in glacial deposits, thought to be caused by melting of stranded remanent block of ice.

Lagerstätte
German word, = stratum + place; sedimentary deposit with notable fossil content, e.g. Conservation-Lagerstätte, with exceptionally well-preserved fossils.

Lahar
Mudflow of volcanic debris, pyroclasts, etc in water; may travel at high speed.

Lamina
Bed thinner than 1 cm (arbitrary division point); laminated, sediments with such character.

Laminated
Lamina.

Landslide
Comparatively rapid gravity-controlled movement of earth material over distinct surface(s) of failure that may be curved or sub-planar.

Lapilli
Material in size range 2-64 mm ejected from volcano.

Laurentia
Former huge landmass comprising much of present-day north America and Greenland, and also north-western Britain and Ireland. Laurentia collided with *Avalonia* and other landmasses in early to mid-Palaeozoic times, forming *Laurussia*, and the British section split away from America from the Jurassic onwards.

Laurussia
Continental grouping of north America, Greenland, and southern and western Europe during Carboniferous times approximately.

Lava
Magma that has erupted at the surface.

Law of Superposition
The principle that in an undisturbed pile of sediment, the rocks at the top must have been deposited on the rocks underneath and must be younger.

Leat
Surface channel for conveying water from source to (in this context, usually mining) site; may be kilometres long; almost parallel to contours. May provide useful clue to past mining activity.

Levee (Fr. *levée*)
Bank of river raised naturally or artificially above floodplain level.

Level (Welsh *lefel*)
Horizontal mining tunnel. Often used to describe an *adit* (or day level) but in Devon/Cornwall, levels never reach the surface; cf. *heading, crosscut, drivage*, road. A drainage level (sough, water gate) provides natural drainage to part of mine above it and discharges to lower ground away from mine.

Limestone
Rock formed mostly of calcium carbonate and variable amounts of magnesium carbonate. Very variable; e.g. includes *oölites*, can be chemically precipitated or formed of biotic debris.

Lithology
The description of the macroscopic characters of a rock, e.g. texture; cf. *petrology*.

Lode
Mineral vein or group of veins. In Cornwall, refers only to workable part of such system.

Longwall
Method of working coal on long faces, in sections termed panels. Suited to favourable geological conditions and mechanised extraction; 'advancing' (moving away from the shaft) or 'retreating' (working back towards the shaft from end of haulage road) types.

Macrocrystalline
Having crystals visible to the unaided eye.

Made ground
Area where land surface has been raised artificially by tipping, backfill of excavation, etc.

Magma
Hot, mobile rock formed by total or partial melting of crustal or mantle rocks; may contain dissolved gases and already-formed *crystals*.

Marble
Metamorphically recrystallised limestone (term often misused by selling agents to refer to a polished limestone); Saccharoidal marble has a sugar-crystal texture.

Marine band
Relatively thin body of sediment produced by short-lived marine transgression into otherwise continental sediments; particularly important for correlation in upper Carboniferous (Silesian).

Marl
Calcareous mudstone, in loose terminology.

Member
Subdivision of *formation*.

Mesozoic
Era of Phanerozoic including Triassic, Jurassic and Cretaceous periods and systems; see illustration of Geological Timescale.

Meta-
Prefix denoting metamorphosed rock, as in metagranite, metasediment.

Metamorphic
Rock formed by recrystallisation of earlier rock *in situ* by heat and/or pressure; e.g. *slate, gneiss*.

Methane
Combustible gas, CH4, released by organic decomposition including from coal; 'firedamp' see *Coal Measure Methane*.

Microcrystalline
Having crystals not visible to the unaided eye.

Mid-ocean ridge
Rift valley and undersea mountain range formed where ocean plate sections separate and move apart by addition of new ocean floor (sea-floor spreading).

Mine
1. Hole or system of holes in the ground from which minerals are extracted through narrow apertures, as opposed to *opencast* mine.
2. *Clay ironstone*.

Mineral
Naturally occurring inorganic substance, usually with recognisable crystalline structure, chemical composition, and other properties.

Mire
Swamp or bog where organic remains accumulate due to diminished decomposition; can give rise to coal deposits.

Mudstone
Rock formed of mud-size sediment particles, usually not fissile.

Mundic
Pyrite; name used mostly in Devon/Cornwall where the pyrite tends to contain considerable arsenic (arsenical pyrite or arsenopyrite); see *Mundic problem*.

Mundic problem
Use of mine waste containing pyrite to make concrete blocks that fail by oxidation of the pyrite. Such blocks are found in properties constructed about 1920-1940, mostly in Devon and Cornwall.

Nappe
Mass of rock thrust on large scale, typically kilometres in lateral extent, over other rock in major tectonism (*see tectonic*). Present in Scotland/Ireland (Caledonian) and Devon/Cornwall (Variscan).

Old Red Sandstone
Old but still-used term for the mostly terrestrial deposits of the British Devonian, with parts of the upper Silurian and lowest Carboniferous. Mostly red or reddish, but not necessarily so, and not necessarily sandstone.

Oölite
Limestone formed of small subspherical grains, like fish roe, such as Bath Stone.

Opencast, open-cast
"Strip mining" where large strips of land are excavated to reveal minerals. Waste is backfilled into worked areas. Often used as synonym for *open-pit*.

Open-pit
More compact, often much deeper excavation; waste is tipped outside. May be used as synonym for opencast.

Openwork
Pit created by extraction of mineral vein from surface; typically irregular in outline.

Orogenesis, orogeny
Orogenesis, mountain-building; orogeny, particular time of mountain-building; orogen, the mass of crust affected by an orogeny. See *Variscan, Caledonian, Alpine*.

Orthogneiss
Gneiss of presumed igneous origin.

Outcrop
The area in which a rock formation comes to the surface (but not necessarily visible; i.e. covered by soil and vegetation); not the same as an exposure.

Outlier
Small area of younger rock completely surrounded by older rock; cf. inlier.

Overhand stoping
Stope.

Palaeo-, paleo- (US)
Prefix from Greek palaios 'ancient', applied to the study of things in the geological past, as in palaeobiology, palaeomagnetism etc. Just add palaeo-, and it's geological.

Palaeogeography
The study of ancient geography, i.e. distribution of landmasses, seas etc.

Palaeosol
Preserved ancient soil, e.g. cornstone, seatearth (under coal).

Palaeozoic
Earliest Era in Phanerozoic, including Cambrian, Ordovician, Silurian, Devonian, Carboniferous and Permian periods and systems; see illustration of Geological Timescale.

Panel
Block of coal to be worked out in *longwall* mining.

Paragneiss
Gneiss of believed sedimentary origin.

Pelite
Metamorphosed mudrocks.

Pericline
Fold that results in subcircular to oval pattern of outcrops; upfolded pericline = dome; downfolded pericline =basin.

Period
Span of Geological time smaller than an *era*.

Petrology
The study of rocks, including composition, mode of occurrence etc; cf. *lithology*.

Phanerozoic
Span of Geological time including the Palaeozoic, Mesozoic and Cenozoic eras; commenced c. 543 million years ago.

Phenocryst
Large crystal surrounded by matrix of smaller crystals; often good crystalline form; thought to have started by slow cooling prior to movement of magma to a location where it cools more rapidly.

Phreatic
Zone below *water table*, saturated with water.

Phyllite
Roughly, intermediate between *slate* and *schist*; characteristic sheen.

Piezometric surface
Notional surface joining levels to which water in confined aquifer rises in observation wells. Cf. water table.

Pillar
Area of unworked mineable ground, of varying size, left to support the roof or important structures such as a mine shaft (the 'shaft pillar') or a power station; or between adjacent mines; cf. *pillar-and-stall*.

Pillar-and-stall
Method of underground working of coal or other material (disposed horizontally or sub-horizontally) in small chambers separated by 'pillars' to support the roof; differing techniques in different areas; also *stoop and room, room and pillar,* bord and pillar.

Pillow lava
Lava formed of accumulation of pillow-shaped lumps, erupted underwater

Pipe
Horizontal to sub-horizontal mineral vein, limited lateral extent or subcircular section; northern British use.

Plate
Plate tectonics.

Plate Tectonics
Notion that most of the earth's gross surface features can be explained by the relative movement of sections of the crust, or plates; 'continental drift'.

Plutonic
Igneous rocks formed at great depth, typically in large bodies; cf. *hypabyssal*.

Portal
Entrance to subhorizontal mine tunnel, or to road, rail or canal tunnel.

Portal
Entrance to adit, etc, at surface.

Pot
Swallow.

Pothole
1. See swallow.
2. Subcircular depression in hard rock on shoreline or river bed, eroded by swirling pebbles.

Prop
Timber or other support to rock (roof) in area of worked-out coal seam or other void.

Psammite
Metamorphosed sand-grade rocks.

Pyrite, Pyrites, Iron Pyrites
See also *mundic*. 'Fool's Gold', Iron Sulphide, with brassy lustre; common mineral in ore veins, shales etc. Ironically, can contain gold in workable quantities.

Pyroclast, pyroclastic
A pyroclast is a small fragment erupted from a volcano; pyroclastic deposits are the result.

Quartz
Highly resistant mineral, silicon dioxide (=silica); usually the major constituent of clasts (=grains) in sandstones.

Radiometric
Radiometric dating permits relatively precise absolute (as opposed to relative) dating of rocks or fossils by comparing the proportions of the included radioactive *isotopes*, which are usually present in trace quantities. There are numerous methods, but all depend upon knowing the half-life of the isotope, that is the time for half of any given mass of the 'parent' isotope to decay to the 'daughter' isotope.
Radon
Gas produced by decay of naturally-occurring radioactive elements, sourced mainly in igneous rock but found secondarily in many limestone areas.
Raised beach
Former beach, now found above present sea level as result of lowering of sea level and/or earth movements; may contain fossils.
Rake
Vertical to subvertical mineral vein; northern British use.
Regional metamorphism
Large-area metamorphism caused often by plate collision and mountain-building.
Regolith
Uncemented, weathered material that rests on unaltered rock, including soil, rock fragments, etc.
Regression
State in which sea level is falling relative to the land.
Resurgence
Location where underground watercourse emerges to daylight.
Reverse fault
Fault of which the hanging wall is pushed up relative to the footwall; opposite of normal fault; cf. *thrust* fault. Result of compression.
Rheic Sea (or Ocean)
Sea that formerly divided *Avalonia* and *Gondwana*; closed by end of Carboniferous.
Rhyolite
Acid eruptive lava, not very common compared with *basalt*.
Rift valley
Linear feature on continent caused by subsidence over line of plate splitting and incipient spreading.
Rise
Mine shaft driven upwards from below ground.
Road, Roadway
Sub-horizontal mining tunnel, usually for transport; tramming road, ditto with tramlines.
Roof
In coal mining, rock above seam.
Room-and-pillar
Pillar-and-stall.
Sandstone
Rock formed of sand-grade sediment, usually cemented to some degree.
Sapropelic coal
See *Coal*.
Schist
Regionally metamorphosed rock with typical laminar texture; more intensely metamorphosed than slate; usually once a mud-grade sediment.
Schistosity
Texture of a *schist*; the alignment of platy and elongated minerals in a planar arrangement, at right angles to the direction of pressure.
Seatearth
Argillaceous palaeosol developed under coal; fireclay. Cf. *ganister*.
Sedimentary
Loose term for rocks that are neither igneous nor metamorphic, mostly deposited in water, occasionally on dry land.
Sedimentary structures
Features of bedded sediments caused by currents, flow oscillations, animal activity, etc. Very useful for environmental interpretation, *way-up criteria*, etc.
Series
Body of sedimentary rock formed during an *epoch*.
Sett
Area of ground entitled to be worked by particular non-coal mine; cf. take.
Shaft
Vertical* mine access; **upcast**, shaft used for expelling spent ventilating air; **downcast**, shaft that takes in fresh air; **winding shaft**, shaft that takes main traffic of men and material; **engine shaft**, shaft with winding engine; **climbing shaft**, shaft for man-climbing. *In Devon and Cornwall, shafts were often not vertical but followed the angle of the *lode*.
Shakehole
Swallow.
Shale
Closely laminated mudstone, usually not very hard, often fissile.
SHRIMP
Sensitive High Resolution Ion Microprobe: precise method of dating *zircon* grains by *radiometric* techniques.
Siderite
Iron carbonate, concentrated in *clay ironstone* to form a workable iron ore.

Siliciclastic
Adjective for *sedimentary* rocks derived by erosion of predominantly silicate-based rocks and minerals.

Sill
Igneous intrusion, generally tabular; more or less in plane of bedding (concordant) or at a slight angle (discordant); relatively thin compared with the sedimentary pile – metres to hundreds of metres thick. Can have great lateral extent (e.g. the Whin sill of northern England); can change level by sudden steps.

Siltstone
Mudrock formed of silt-grade sediment.

Sinistral
Of a strike-slip fault: movement of the far side of the fault to the left of the viewer.

Sinkhole (Welsh *sinc*)
See *swallow*.

Slate
Low-grade regionally metamorphosed rock, highly fissile, with slaty *cleavage*; indurated and cleaved *mudrock* or fine-grained igneous rock; used for roofing and (formerly) damp-proof courses; originally usually mud-grade sediment or fine-grained *volcaniclastics*.

Slaty cleavage
The type of metamorphic cleavage that forms slate.

Sole marks
Characteristic small erosional features, formed by currents, that may be preserved as sediments later fill them.

Solifluction
Gravity-driven flow of saturated soil, other superficial deposits etc.

Sough
Drainage *level* or *adit, water gate*; provides natural drainage to part of mine above it and discharges to lower ground away from mine.

Sphalerite
Main ore of zinc, zinc sulphide; also Blende, Black Jack.

Spreading
Separation of plates, usually along *mid-ocean ridge* or *rift valley*; opposite process to *subduction*.

Stage
Body of sedimentary rock formed during an *age*; divided into *zones*.

Staging
Platform erected in shaft or other void to allow convenient access to working area above local floor level.

Stemple
Timbers wedged into shaft or other void, used for climbing; may also support staging for working.

Stoop and Room
See *pillar-and-stall*.

Stope
Extraction chamber in mine (generally, for metal ore); **overhand stoping**, extracting upwards from gallery; **underhand stoping**, extracting downwards from gallery.

Stoping
Stope.

Stratigraphy
Study of sedimentary or metamorphosed sedimentary rocks, especially their history, disposition and correlation.

Stratum
Plural strata; Latin =layer; rocks that form beds, but no specific inference as to age, thickness or extent. A useful loose term.

Strike
Notional line at right angles to *dip* but in plane of earth's surface (assumed flat) e.g. strike valley: valley approximately at right angles to dip.

Strike-slip fault
Transcurrent, or tear, fault; fault with more lateral than vertical movement.

Stull
Timber or other prop inserted to provide support to opposite sides of worked-out vein or other area; mostly Devon/Cornwall usage; cf. *prop*.

Stylolite
Irregular solution surface within carbonate rock, usually marked by thin insoluble residue.

Subcrop
The area in which a rock formation abuts the underside of an unconformable later formation.

Subduction
Process involving sinking of a crustal plate under another where they meet; opposite process to *spreading*.

Superficial deposits
Unconsolidated *alluvium*, *till*, or other materials, generally of Recent origin and overlying older, more consolidated rocks. Referred to as 'drift' on many maps.

Supergroup
Lumping together of adjacent *groups* of rock *formations*, as in Dalradian Supergroup.

Superposition
See *Law of Superposition*.

Suture
Long, generally narrow feature of highly deformed rocks and often suspected ocean-floor fragments, interpreted as junction line of collided continental masses or *terranes*.
Swallow, Swallet, Swallow-Hole
Enclosed depression in limestone (karst) area where watercourses have sunk into the limestone to emerge elsewhere; may lead to extensive cave system. Known as doline in original Karst area; also (in UK) *shakehole, pot* or *pothole*.
Syenite
Intermediate deep-seated equivalent of *trachyte*.
Syncline
Downfold in sedimentary strata; cf. *synform*.
Synform
Looks like a *syncline* but may include inverted strata.
System
Body of sedimentary rock formed during a *period*.
Tailings
Usually fine-grained material left at end of mineral processing, not worth further processing and left in dumps that may give rise to an environmental hazard.
Take
Area of ground worked, or entitled to be worked, by a particular coal mine; cf. *sett*.
Taxonomy
The study of the ordered arrangement and classification of individual types of life forms.
Tectonic
Term relating to deformation of the earth's crust (tectonism) and its subsequent structural effects.
Terrace
River terrace, area of abandoned floodplain alluvium left at higher level as the river and its floodplain erode downwards.
Terrane
Area with characteristic geology, usually separated from adjacent terranes by major faults or suture lines; a continental fragment.
Thrust fault
Fault where one side is pushed over the other, usually at a low angle; result of compression.
Till, tillite
Till is material laid down by a glacier without any involvement by water. Typically, unstratified, unsorted, and containing a wide mixture of material derived from the land over which the glacier passed. Tillite is an indurated till. Cf. *boulder clay, diamicton*.
Toadstone
Lead miners' term for igneous rocks found within mine.
Trace fossil
Footprint, burrow, feeding marks or other impressions left in rocks by organisms.
Trachyte
Intermediate eruptive lava.
Transgression
State in which sea level is rising relative to the land.
Travertine
Calcium carbonate deposited from carbonate-saturated waters, notably at hot springs; often in caves; often shows concentric internal structure with small voids. Cut and used for floor and wall finishes.
Trilobite
Common Palaeozoic arthropod, vaguely similar to woodlouse but not closely related.
Tuff
Lithified volcanic 'ash' deposit (air-fall or water-laid), with pyroclasts less than 2 mm. Over 2 mm size are *lapilli*, hence lapilli-tuffs.
Tuffaceous
Sediment with significant admixture of *tuff*.
Turbidite
Sediment deposited by *turbidity current*.
Turbidity current
Current with load of sediment suspended by internal turbulence; tends to produce internally graded bedding, usually fining upwards as flow slackens. Product, *turbidite*, common in British *Palaeozoic* rocks.
Turn
Winze.
Unconformity
Gap in sedimentary rock succession formed by non-deposition or erosion of strata; several different types distinguished. Most conspicuous type is angular unconformity where the later rocks lie across the tilted and eroded edges of the strata below.
Underground Coal Gasification (UCG)
Method of combusting coal at depth to extract useful fuel gases. Could be environmentally disastrous.
Underhand stoping
Stope.
Unfossiliferous
Bearing no fossils; opposite of *fossiliferous*.
Vadose
Zone above *water table*, not completely saturated with water.

Valley bulge
Distortion of bedding and dip in valley bottom eroded into clays or other soft strata, caused by plastic flow.

Variscan
Term given to period of major earth movements (*orogeny*) that gave rise to massive mountain ranges running roughly west to east across central Europe; responsible for major structures of southern Ireland, south Wales, Bristol area, and Devon and Cornwall; late Devonian to late Carboniferous. Caused by collision of Gondwana with Laurussia.

Variscan Front
Approximate boundary of major deformation by Variscan *orogeny*, running through southern Eire, south Wales, the Mendips and (at depth) southern England towards Dover.

Volcanic breccia
Volcaniclastic sediment formed of blocks ejected from volcano, over 64mm in diameter.

Volcaniclastic
Rock formed of fragments generated by volcanic action.

Water gate, water-gate
Drainage *level* or *adit; sough*.

Water table
Boundary (a notional surface) between *vadose* and *phreatic* zones. Not fixed, and not necessarily level or smooth. Below the water table, water is at greater than atmospheric pressure and can emerge as springs where the water table intersects the ground surface. Cf. *piezometric surface*.

Way-up criterion
Sedimentary or other feature that forms with distinct top and bottom and can therefore be used to infer overturning of strata; e.g. *sole marks*.

Welded tuff
Obsolete term for *ignimbrite*.

Wheal
Devon/Cornwall term for mine (but includes *opencast*, as in china clay pits).

Whinstone
Basalt.

Winding shaft
Shaft.

Winterbourne
Seasonal stream that flows only after persistent rain, usually in winter. Common in Chalk areas.

Winze
Mine shaft that starts underground from tunnel or chamber; also called *turn*.

Zircon
Zirconium silicate, accessory (i.e. very minor) mineral in igneous rocks; forms squarish crystals (diamond substitute in jewellery). Also found in gneisses and schists. Hard and resistant to erosion and chemical alteration, and sought for radiometric dating purposes. Unaltered parts of the zircon can be dated with considerable precision by *SHRIMP* methods.

Zone
Small unit of sedimentary rock characterised by particular fossil species; used for long-distance time correlation. Can be grouped into *stages*.

Index